Teaching 21st Century Skills

Rekha B. Koul · Rachel Sheffield ·
Leonie McIlvenny

Teaching 21st Century Skills

Using STEM Makerspaces

Springer

Rekha B. Koul
Curtin University
Perth, WA, Australia

Rachel Sheffield
Curtin University
Perth, WA, Australia

Leonie McIlvenny
Curtin University
Perth, WA, Australia

ISBN 978-981-16-4360-6 ISBN 978-981-16-4361-3 (eBook)
https://doi.org/10.1007/978-981-16-4361-3

This Springer imprint is published by the registered company Springer Nature Singapore Pte Ltd.
The registered company address is: 152 Beach Road, #21-01/04 Gateway East, Singapore 189721, Singapore

Foreword

The OCED and a range of employers have on many occasions described the need for emerging employees to be able to demonstrate collaboration, communication, critical and creative thinking and resilience. The Gonski report and the Melbourne Declaration also indicated that students needed to be upskilled in twenty-first-century skills or transversal competencies. What has not been forthcoming until now has been a guide to help teachers identify, facilitate, and then assess these skills. This book meets this need and provides teachers with a range of strategies, tools, and ideas to help them to embed these skills into their practice. As experienced educators, the authors are aware that a Makerspace offers an ideal opportunity for teachers to facilitate the development of these skills in their students, while complimenting the current school learning frameworks.

Congratulations go to the team for a useful and practical book that provides examples that link to curricula and tentatively explore the future for education with Education 4.0.

North Perth, WA, Australia

Geoff Quinton
Australian Science Teacher Association President
(2017–2019)

Acknowledgements

This book could not have been completed without the diligence energy and effort of a number of key contributors.

These include Leonie McIlvenny who spent many many hours on the reflection around these key ideas and with whom we had many interesting and enlightening conversations. Joel McIlvenny who was an excellent copy editor ensured that the book was consistent and well written, and we really appreciated his help and support.

The iSTEM student teachers in the 2019 and 2020 cohorts worked with school students and bought their thinking and their experiences and shared them with us to include in the text. We thank two colleagues and excellent educators, Michael and Nicole, for sharing their experiences and their writing about their schools and classrooms.

2021

Rekha B. Koul
Rachel Sheffield

Contents

About the Authors

Rekha B. Koul is Associate Professor at STEM Research Group, School of Education and Dean International in the Faculty of Humanities Curtin University, Perth, Australia. Her expertise lies in the development, refinement and validation of questionnaires; investigations of the effects of classroom environments on student outcomes; evaluation of educational programs; teacher action research aimed at improving their environments and evaluation of curriculum. Her publication record includes one authored and seven edited volumes, four book chapters and many journal articles published in peer-reviewed journals.

Rachel Sheffield is Associate Professor in the School of Education at Curtin University, Australia, and is a passionate science educator. She researches and publishes in science and STEM education and professional identity and is currently exploring the transversal competencies and their role in STEM. Rachel has won several faculty, university and national awards for teaching excellence and was awarded an Executive Endeavour Fellowship in 2016. She is also Chair of the prestigious Curtin Academy.

Leonie McIlvenny is an educator and curriculum consultant who has worked across all educational systems and sectors in Western Australia. She currently works as Sessional Lecturer and Research Assistant at Curtin University, Perth, Australia. She researches and publishes in the areas of school libraries, information literacy, digital literacy and 21st-century (transversal) competencies. Leonie has written and published numerous articles around these topics and has presented at local, national and international conferences.

Part I
About Makerspaces

Preface

Part I will provide information around the history of and rationale for Makerspaces and examine a number of locations where people are encouraged to design and make. Part I explains the evolution of Makerspaces and considers the variety of Makerspaces globally and their purpose, intent, and how they engage with their local community.

Chapter 1. "The Story of Makerspace"

This chapter considers the evolution of Makerspace from Hackerspace and the diversity of these spaces from hobbyist corners in libraries for knitting and craft to high-tech spaces where robotics and coding are the norm. Makerspaces exist on a continuum between the free form of traditional Makerspaces and the more rigid structure of Makerspaces in classrooms. Makerspace quadrant model was designed to show various approaches when considering a Makerspace.

Chapter 2. "Makerspaces and STEM—A Synergistic Partnership"

This chapter explains the history of STEM as a relatively new construct that focuses on the synergistic relationship between the areas of science, technology, engineering, and mathematics. This chapter brings together the needs of school and industry to create STEM Makerspaces that seek to bring together skills and concepts under a new STEM theoretical framework.

Chapter 3. "The World of Makerspaces"

This chapter examines Makerspaces globally and identifies their purpose, contexts, audience, and goals. It will also describe how Makerspaces have been created and the strategies employed to allow them to remain relevant. Examples include: the Malaysia shopping centre Makerspace in Kuala Lumpur, Questacon in Canberra, Australia, and the Science Centre Makerspace in Pune. It considers Makerspaces not just from the makers perspectives, which in the case of schools is the students' focus, but also as conceptual pedagogical play areas for teachers to explore their practice in a safe space where failure is encouraged.

Chapter 1
The Story of Makerspace

Making can be a messy activity – this can be confronting to colleagues and school leaders.
(Michael Graffin, 2020)

Keywords Makerspace · Design thinking · Tinkering · Creativity · Construction · Making

Definition of a Makerspace
A successful, sustainable **school Makerspace** is grounded in a clear vision of purpose and pedagogy, develops the capacity of the participants (students and teachers), encourages exploration and play, and provides an openness to risk-taking and failure.

Focus Questions

- What is a Makerspace?
- How did Makerspaces develop and evolve?
- What are the essential elements of a Makerspace?
- What is *Makerspace Quadrant Model* and how does it support maker creation?

R. B. Koul et al., *Teaching 21st Century Skills*,
https://doi.org/10.1007/978-981-16-4361-3_1

1.1 The Beginning of Makerspaces

There is a view that Makerspaces are new designs that have only existed since the creation of the online Hackerspace (Copyright© 2016 London Hackspace Ltd.). While the term is relatively new, the ideas of making and constructing have been developed over generations. It could be argued that the concept of a Makerspace was conceived at the time of Hailmann, Dewey, or Montessori, when the concepts of problem-based learning, project-based learning, constructivism, and constructionism became popular (Bilkstein, 2018). Dewey saw the development of individual understanding as an agile reaction to our surroundings, arguing that learning 'cannot take place by direct conveyance of beliefs, emotions and knowledge… it takes place through the intermediary of the environment' (1916). Dewey's notion of experience is 'broadly conceived… [and] more than simply a matter of direct participation in events' (Rodgers, 2002). He saw learning as 'the self-controlled or directed transformation of an indeterminate situation into one that is so determinate in its constituent distinctions and relations as to convert the elements of the original situation into a unified whole' (1938).

Mathematician Seymour Papert built on the work of Piaget, suggesting that learners be inventive and build artefacts including robots. This supports a constructivist approach to learning. It was also Papert who elevated the cognitive status of making by looking at the complexity of a design, moving from the abstract to the concrete, and then from the concrete to the abstract. What initially seemed a frivolous corner of a technical drawing class where art meets production was actually based around some of the most highly regarded pedagogy of the century?

Recent History

The maker movement and the term 'maker' gained traction through Maker Fairs and FabLabs. They developed organically from the online Hackspace to an actual physical setting. Fabrication Laboratories, FabLabs, are spaces where makers gather to share, exchange, collaborate, and build. Common designs are used to digitally manufacture custom objects, while consistent software and hardware capabilities among different labs allow the distribution of projects across them. They are also commonly established in a more formalised and internal setting compared to many Makerspaces, such as a company or university. Each FabLab is also supported by the global FabLab Association, which drives promotion and support of the FabLab concept and collaborations between different FabLabs (Rosa et al., 2017).

Hackerspaces function as community labs and involve people interested in electronics and programming who follow hacker and maker practices. These participants, spaces, and communities are of increasing interest because they offer areas for interactions between people and computer technologies (Toombs, 2017).

Termed a 'place for making' or 'Makerspace' (Smith, Hielscher, Dickel, Söderberg, & van Oost, 2013), these settings provided new opportunities for makers to create any unique item that their imagination and skills allow, thus engendering exciting and engaging spaces to experiment and play. They have also been considered trendy new settings where people can go and create on their own or collaborate with others on a shared idea. Unlike a factory where all items are the same, these items were unique and belonged solely to the individual.

The evolution of Makerspace has resulted in many variations in its definition, with some of the more interesting terms using language to capture the essence of each space. For example, Peppler and Bender (2013) point to: 'a growing culture of hands-on making, creating, designing, and innovating… [and] a do-it-yourself (or do-it-with-others) mindset that brings together individuals… making nearly anything' (p. 23). Table 1.1 provides definitions of Makerspaces according to different researchers and academics.

Table 1.1 Definitions of a Makerspace

Author	Quotes defining a Makerspace
Kurti (2014)	The ideal space for maker education
Sheridan, et. al. (2014)	Using creative STEM skills combined with art in a physical space which explore ideas and develop new products and learn technical skills
Oliver (2016)	Technical projects pursued by individuals in a shared physical space which has the affordances of a maker community
Bowler and Champagne (2016)	A place where technology industrial and fine arts are explored through the work of a hands-on collaborative community
Cohen (2017)	Where physical artefacts are constructed, deconstructed, and reconstructed shared with a community of makers
Freeman, Becker and Cummins (2017)	Integrating the maker mindset into the formal curriculum to spur real-world learning
Giannakos, Divitini and Iversen (2017)	Making to invent new forms of expressiveness and utilise technology to support twenty-first-century education
Halverson and Sheridan, (2014)	Makerspaces hold collective activities that seek the construction and diffusion of alternative forms of innovation and collaboration, rooted in the thinking of the maker movement
Taylor, (2016)	The maker movement is an extension of the do it yourself (DIY) culture based on practices of creating, building, modifying, and repairing something using traditional or digital tools and machines

1.2 Makerspace Versus Maker Program

Educators need to carefully consider the purpose and reason for their Makerspace to ensure its efficacy for student success. Within a Makerspace, the learning that occurs could be placed on a continuum. At one end of this continuum, learning is highly structured with specific outcomes (intentional learning), and at the other, learning is a more organic process driven by the learner where there are no set outcomes (incidental learning).

$$\text{Intentional Learning} \leftrightarrow \text{Incidental Learning}$$

The making space itself might also be considered on a continuum where at one end Makerspace hosts a very structured maker program while at the other, it is a more flexible space that changes constantly. One is driven by the program itself, the other is driven by the interests of the makers.

$$\text{Makerspace} \leftrightarrow \text{Maker Program}$$

A maker program or maker approach suggests that making can occur organically in any appropriate space. This may look like activities conducted in a library, community centre, tech-lab, or any other suitable environment. The direction or focus of the making as it relates to an idea. Table 1.2 describes some of the differences between a Makerspace and a Makerspace program.

Table 1.2 Makerspace Versus maker program (Blackley et al., 2018)

Traditional Makerspace	Makerspace program: Targeted learning approach
Makers create their own communities	Makers are organised into predetermined communities
Makers choose materials at their own discretion	Makers are provided with a base-level kit of materials
Makers envisage and produce individual, often unique, artefacts	Makers are shown a completed base-level and operational (as appropriate) artefact and are challenged to construct a similar artefact
Makers are not mentored	Makers are mentored (not instructed)
Makers might evaluate their artefact	Makers are guided through scaffolded criteria to evaluate their artefact
Makers might be cognisant of underlying science, technology, engineering, mathematics or other concepts; however, this is incidental to the activity and experience of the maker	Makers are made aware of related underlying science, technology, engineering, mathematics, or other concepts in line with curriculum documents

Tables 1.1 and 1.2 illustrate how the maker movement has evolved organically in different communities to meet the needs of the community naming it. For example, Makerspaces can be creative knitting corners in nursing homes, multipurpose, and diverse rooms in a library, collaborative industrial areas where artists can share machinery, high tech centres in big companies, and online hacking spaces for programmers and gamers to meet and generate ideas and innovate new products.

All Makerspaces occur within a space that provides resources and opportunities that allow a collective or community to make an artefact or product that is often unique to the maker yet can be based on a common theme and even a common pattern (Shively, 2021). When considering other aspects relevant to Makerspace community, it is important to widen its definition to ensure inclusivity of the unique community aspect. When creating a Makerspace, the design and purpose must meet the needs of the community, and keeping the design, resources and pedagogies 'open' will ensure that those needs are met.

Makerspace Quadrant Model

Makerspace Quadrant Model describes Makerspaces within four contexts that accommodate both the intentional/incidental and Makerspace/maker program continuums. This model not only allows educators to identify what they want their Makerspace to look like, but also what form the learning within will take (i.e. the learning theories applied, and pedagogical approaches taken). Figure 1.1 describes how a Makerspace can be created and what it might look like.

Makerspace Quadrant Model

Within each quadrant, there are a number of features or characteristics that define it and separate it from the other quadrants. These are described in Table 1.3—Makerspace Quadrant Model.

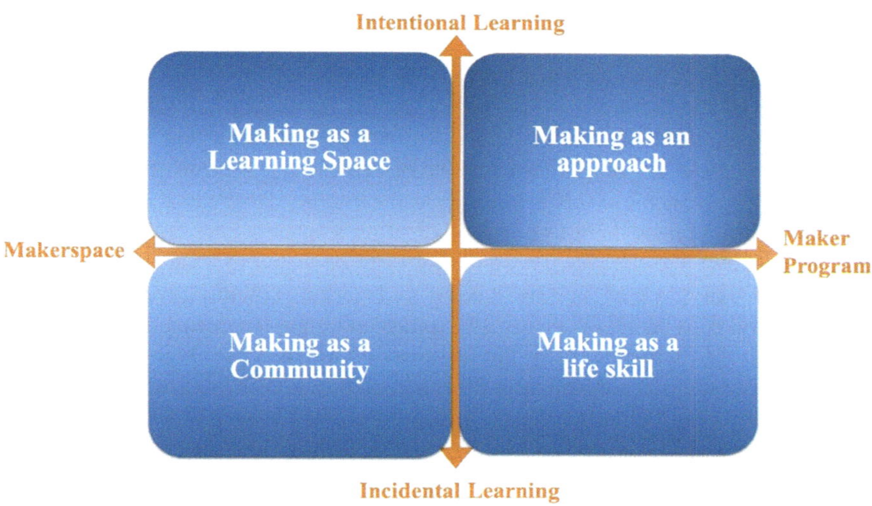

Fig. 1.1 Makerspace Quadrant Model

Table 1.3 Makerspace Quadrant Model

Making as a learning space	Making as an approach
The space itself (along with the resources) drives or influences the learning Woodworking lathes, pottery wheels, 3D printers, and electronics equipment will all influence the type of undertaken activities	The learning that occurs in this space is driven by the learning philosophy and theories that drive the activities. This may include constructivism, design thinking, prototyping, and collaborative problem-solving
Making as a learning community	Making as a life skill
Community emphasis will impact the various demographics of the group These groups may develop organically out of interests, strategically created by balancing skills and abilities or some other external driver (ability groupings in an educational setting)	The emphasis within Makerspace is the skills set that is developed through undertaking a maker project. Skills might include communication, collaboration, critical thinking, coding, and project management. The emphasis is the development of the individual

1.3 Makerspaces and Maslow's Hierarchy of Needs

In 1943, American Psychologist Abraham Maslow created two hierarchies of need: The *Hierarchy of Basic Needs* that described a range of biologically rooted needs and the *Hierarchy of Cognitive Needs*—the need to know and understand. Maslow believed that these were essential elements that individuals sought throughout their life. If these elements are as important to the individual, as Maslow suggests, then it could be extrapolated that the hierarchy's application to the classroom, or in this case, Makerspace is relevant and useful as a way of preparing students for their future life in the world of work and adulthood. In this section, we will examine how Makerspace can support and nurture the individual through each of the stages of the hierarchy. In Section Two, Maslow's hierarchies will also be explored in terms of the targeted competencies examined in that section.

(Maslow, 1954, Motivation and Personality, p. xii-xiii).

The apex of the hierarchy is self-actualisation, which Maslow saw as realising personal potential, self-fulfilment, personal growth, and seeking peak experiences; a desire 'to become everything one is capable of becoming' (Maslow, 1987, p. 64). If we want children to achieve self-actualisation, then we need to create the environment and the opportunity for them to do so by creating a space where children's beliefs about themselves and the world around them can be tested in a safe and nurturing environment. This place of 'safe risk-taking' and working collaboratively with others to seek solutions to problems or just to explore and create are often central themes that influence the nature and drive the features of a Makerspace (Fig. 1.2).

Considering the mechanisms of a Makerspace beside Maslow's hierarchy, there is capacity to apply the hierarchy in a Makerspace environment for developing a student's twenty-first-century skills and overall learning capabilities.

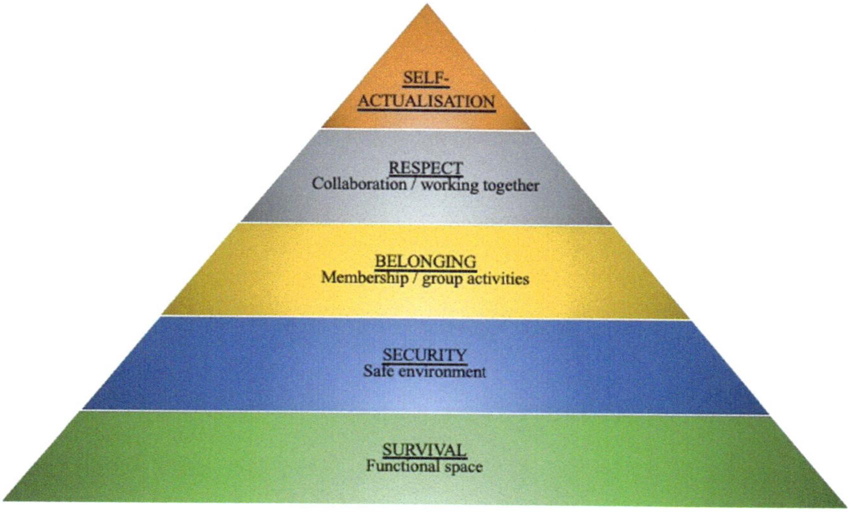

Fig. 1.2 Makerspaces within Maslow's hierarchy

Table 1.4 considers how a maker would progress through the respective stages of Maslow's hierarchy within a Makerspace environment. Initially, the individual would start in the survival stage (in a functional space) and eventually establish feelings of safety and security, followed by the comfortability to participate in group activities. This would eventuate in engagement with collaborative projects in Makerspace, followed by empowered self-directed learning.

The community in the Belonging (Maslow) or with a Makerspace membership (Maker) starts with the group identity, branding, outreach, social media and moving through and towards intergenerational skills, social justice alliances at the highest level of transcendence and sustainability the highest level of the maker hierarchy. A maker mindset or *maker culture* is a way of thinking that includes:

- a hands-on approach;
- an enthusiastic 'can do' attitude that demonstrates a problem-solving capacity to overcome roadblocks and issues;
- encouraging resilience and risk-taking skills to overcome failure and pushing boundaries when not successful;
- the capacity to be a lifelong learner;
- flexibility and adaptability when working with materials or space that is sourced or available at the time;
- supporting makers to effectively work collaboratively.

Table 1.4 Maslow's hierarchy within Makerspace

Maslow hierarchy /Maker hierarchy	Space	Maker	Curriculum	Community
Transcendence/ sustainability	Regional networks, conferences, publications	Transformational leadership, advocacy, mentorship	Research, new technologies, ongoing projects	Donors, intergenerational skills, social justice alliances
Self-actualisation/ results	maker in residence, digital badging, in-house internships, innovation	Program completion, employment, self-agency, entrepreneurship	Real-world problems, assessment portfolio, micro-enterprises	Recognition, partnerships, internships K-12
Respect/creative collaborations	Collaborative spaces, project storage area	Growth mindset, experimentation, skill development, mentoring	Interdisciplinary, project based, design thinking, soft skills	Team challenges networking workshops employer outreach
Belonging/Makerspace membership	Event space, welcome sign, member system	Membership, rules, ethics, code of conduct	Diversity, equity, guided orientation, soft skills	Group identity, branding, outreach, social media
Security/safe eenvironment	Visibility, monitoring alarms, egress, signage	Safety training, protective apparel, secure storage	Orientation, safety training, fire drills, first aid, situational awareness	Mentors Safety officers Pastoral care
Survival/functional space	Ventilation, lighting, toilets and sink	Hydration, appropriate clothing	-	-

1.4 Makerspaces in a School Setting

There was initial resistance to the implementation of Makerspaces in schools as there were no perceived links to the achievement of curriculum outcomes. This was due to the open-ended nature of Makerspace activities which implied that teachers could not control the learning. The belief that the curriculum could not adequately be addressed labelled Makerspaces as an extravagant use of limited space. This began to change, however, due to the growing evidence that Makerspaces were based on well-founded ideology, including the deep thinking and knowing of Dewey, Wenger, Freire, Papert, and Piaget.

More recently, Makerspaces have been included in the annual Horizon Report K-12 since 2015, showing the rise of Makerspaces 'from compelling phenomenon to global movement' (p. 40). They have also increasingly been heralded as providing opportunities for learners to engage in creative, higher-order problem-solving through hands-on design, construction, and iteration (Sarkadi, 2020). Employers were also seeking graduates that were versatile and agile thinkers who could think critically and creatively and were able to solve problems. They wanted employees who were ready for a more rapidly changing world (United Nations Educational Scientific and Cultural Organisation, 2017). The Foundation for Young Australians (FYA) has also found that:

> employers have listed more 21st century skills in their job advertisements… [and] the proportion of job advertisements that demand critical thinking has increased by 158%, creativity by 65%, presentation skills by 25%, and team work by 19%" (2016).

The term twenty-first-century skills describe skills such as creativity, critical thinking, communication, and collaboration, which are also known as soft skills (see P21, 2009). The Crockett (2011) developed twenty-first-century fluencies including collaboration and creativity with technology. It is posed that these skills, like language, need to be developed to the point of fluency in the current digital age—refer to Chap. 4—Transversal Competencies.

Based on the research of Papert, Piaget, and Dewey, we believe that a pedagogical approach involving a balance of explicit instruction and open-ended inquiry creates an effective learning environment. Our suggestion to you as an educator is to consider the purpose of your Makerspace. Visualising and setting up the space with purpose and intent is extremely important. Use the next sections *Making a Makerspace* and *Making and Maslow* to consider where to start in the creation and philosophy of your Makerspace. Balancing this clear adhesion to process comes with a warning from Crichton and Carter (2015) about the control through cookie cutter activities:

> Unless educators intentionally pursue innovation and creativity as learning outcomes, Makerspaces will become "imagination ghettos" where issues of access, purpose, and ownership resemble those common in the cloistered environments of early computer labs… students are tasked with cookie cutter activities and trivial projects to complete (p. 3).

Managing Makerspaces in Schools

Planning group-making and collaboration is a difficult challenge that requires a teacher to use strong interpersonal and organisational skills in both the face-to-face and online environment. For educators, these skills need to be considered and refined as a Makerspace may be more unstructured and therefore seem disorderly as it is often noisier with children moving around completing different tasks. Based on Martin's (2015) ideals, there are several fundamentals to developing a prosperous maker culture at school that include:

- supporting professional development to ensure teachers have the necessary skills and confidence to start their journey;
- ensuring the support of school leadership but also the technicians and cleaners you need their support to create Makerspace and sustain the approach;
- focusing on students developing a hands-on inquiry-based approach;
- having resilience to overcome failure and setbacks;

- encouraging risk-taking to help students overcome failure and push boundaries when unsuccessful;
- the capacity to be a lifelong learner with enough energy to continue to learn, search, collate, share, and organise experiences for students;
- the ability to deal with noise and movement in a STEM learning space as students will be working on projects and that can be noisy;
- preparation to not always be an expert and accept the skills/knowledge brought into the classroom/Makerspace;
- supporting effective collaboration.

Some students may already be exceptional 'makers' due to previous engagement in Makerspace type activities in their homes and communities. It is important to embrace this speciality knowledge and encourage these expert makers to support others. Their expertise may also help you to learn skills as you demonstrate that you too are a 'learner' that is keen to play and discover. When carefully constructed and thoughtfully designed, Makerspaces can help develop integral developmental skills that could include the following;

- creativity;
- problem-solving;
- critical thinking;
- inquiry skills;
- design thinking;
- collaboration;
- autonomy;
- literacy;
- numeracy;
- scientific content knowledge;
- maths skills and content knowledge;
- technological knowledge;
- communication skills;
- reflective practices;
- resilience;
- self-efficacy.

Educators need to decide whether student experiences in Makerspace are to be more organic or if there is a particular set of skills or area of content that needs to be taught. The latter of these aligns with a specific maker program rather than the less structured notion of a free form Makerspace. When you do create your own Makerspace, ensure you define it for your colleagues so that there is a shared understanding.

Planning is important for both maker and teacher. The design of the space being used and how materials are included and arranged effects how students engage with their surroundings. For example, a design or model can be expressed on the wall to remind students of Makerspace processes and their possible order.

If the maker is new and inexperienced in making and therefore lacks some of the key safety understandings, an explicit approach will be most useful in quickly bridging this skill gap.

Sustaining Makerspaces in Schools

Teachers can use Makerspaces to create the structure and time for staff to directly collaborate, teach, and reflect on the making process. Teachers are encouraged to become involved, as it is an opportunity for arts, technology, science, and maths specialists to collaborate with classroom teachers in creating curriculum. This may result in STEM projects that embrace art and global projects and consider a wider focus around world problems and solutions.

The capacity for collaboration and inclusion of teams of teachers will determine the sustainability of a Makerspace. If there is too much ownership invested by only one teacher and that teacher leaves, then the entire project and Makerspace may be neglected and deserted. Learning would be supported by technicians who embrace and model the maker mindset and so become important in the sustainability of Makerspace. Parents can also be makers and designers. Direct involvement from kids to ask their parents at home or directly in school /virtual connections will also enable parents—potentially those with manual skills in manufacture and fabrication—to become valued members of the school community.

Materials and Tools in the School Makerspace

Creating a successful Makerspace requires more than having the necessary space and tools: it is also dependent on people who have a 'maker mindset' which encourages risk-taking, thinking outside the box, and seeking solutions to problems. Even the most modern building and the latest technologies do not guarantee a dynamic Makerspace—machines will languish and become dusty when unused by a lacklustre group of or students or makers. There is a benefit to smaller spaces where dynamic personalities manage an initially tiny uninspiring space, and then create a vibrant and active community.

The tools and resources found within Makerspace are still extremely important in determining the depth and breadth of activities that can be undertaken. Some tools are considered essential to a Makerspace, while others reflect specialists' projects and products. It is recommended to start by sourcing free consumable items and cheaper tools initially and then determining what is used most frequently. Proceeding slowly is important and aligning resources with the goals and objectives of Makerspace is essential. Only wasted potential lies in a Makerspace where large shiny exciting items, such as a 3D printer, idly sit.

Equipment Basics

There are a number of simple items that are easy to source and helpful for your Makerspace. These items are listed as: consumables, tools, and both small and large specialist items.

Consumables

The majority of these items are listed in the appendix, and this is by no means an exhaustive or even complete list. You will also certainly add to the items as you create your own Makerspace.

- cardboard;
- butchers paper;
- small white boards;
- paper cups;
- cardboard items (rolls, boxes, etc.);
- *Makedo* cardboard connector kits;
- wool;
- pipe cleaners;
- colouring books;
- origami paper;
- crayons, markers, coloured pencils, pens;
- duct tape, glue, fasteners;
- wood off-cuts;
- fabric off-cuts;
- metal sheets;
- nails and screws.

Tools

- scissors;
- vinyl cutter;
- glue gun;
- soldering iron;
- saw;
- hammer;
- rule;
- safety glasses;
- sewing supplies;
- gloves and protective hand wear.

Extreme care needs to be considered when these tools are handled both by teachers and students. All occupational health and safety guidelines must be adhered to. Makerspace safety needs to be taught to students, and even early primary classroom students can be encouraged and reminded to wear the appropriate safety equipment. In some cases, the children enjoy the dressing up as much as they enjoy using the hammers and nails and seeking help from parents with experience. The wide range of materials also helps engage parents to feel comfortable in the space. Although this can be time consuming, providing parents with a space and opportunities to engage positively in the classroom and ensuring the learning space is a fun place to be will then improve students' attitude and engagement in school.

Specialist Items

- Cubelets, littleBits, ozobots, BBC Microbit;
- snap circuits;
- squishy circuits;
- K'NEX;
- LEGO® (Technic and classic);
- batteries and LEDs;
- math magnetic construction kits;
- motors, batteries, diodes, LED, switches, and wires.

Large Specialist Items

- 3D printer;
- poster printer;
- Arduinos®;
- Makey Makey®.

It may be sensible to use some small items such as motors and batteries to demonstrate how they work before you spend funds on more expensive kits that can be 'black boxes' which are difficult for you to use and then sit unused.

1.5 Teacher Reflection

Before deciding to create a Makerspace and the form it should take, teachers may consider the following questions to ensure the efficacy of their potential Makerspace.

What can I do to support my students' learning?
Learning in Makerspaces is supported by the *place*, the *pedagogical approach,* and the *maker* (the student). You, as the educator, must decide these aspects depending on the purpose of your Makerspace and where it fits into the learning of your students. We imagine that a Makerspace will provide you and your students with a 'play' space where you can experiment with your pedagogical approaches and students can try and fail and review their approaches without fear of permanent impact on their academic record.

Place Design of the space(s) being used and how materials are included and arranged. Schools should be prepared to gift a space to enable a Makerspace to be developed but also to help create the necessary culture.

Pedagogy There needs to be a balance between explicit instruction and open-ended inquiry. This also refers to the general pedagogical strategies that are used, the types of tasks that you set as the teacher, and the sequencing of tasks.

Maker Makers' background knowledge and their ability to collaborate productively. A Makerspace is defined by the talents, interests and abilities of the individuals within it.

Supporting teachers' capacity to work with their students in a Makerspace
Teachers need to be provided with time to learn and play with reliable tools and technology and provided with resources and collegial support. Teachers that are well-organised, pedagogically grounded, and hands-on are better equipped to teach in Makerspaces and to develop the required skills and knowledge. School culture through the principal and the school administration must encourage exploration and give time to experimentation for teachers and students. The school must also support the teachers' participation in professional learning to help them upskill and develop new pedagogical approaches.

What can I do to upskill myself to be confident and competent in the maker area?
Recent research and discussions with students have reported that there are explicit outcomes for students who are participating in Makerspaces. The issue with this research is that it is not around the more regular reflection and assessment that is done in classrooms where the focus is product orientated or on the content rather than the skills.

What and how do I want my students to learn?
What supports learning in maker activities? Learning in Makerspaces is supported by the place, the pedagogical approach, and the maker: the student. There must be a balance between explicit instruction, open-ended inquiry, the general pedagogical strategies used, the types of tasks set by the teacher, as well as the sequencing of tasks. Familiarity with engineering design/rapid prototyping/design thinking are also a complementary part of a maker mindset for teachers and learners.

1.6 What Do Makerspaces Sound like?

Student and Teacher voices in a Makerspace

Creativity and imagination

- I can make anything (student).
- Wow did I make that? (student)
- That was unexpected and it is amazing (student and teacher).
- My girls loved the building challenges, while the boys were excited by creative activities (teacher).

Critical thinking and problem-solving

- I think I can make an amendment to this to make it work (student).
- The students made it and then printed it but it did not work so they went back and had to rework it and then they reprinted it, it was in the review and rework where the major learning occurred (teacher).

Development of content knowledge relevant to the real world

- I can see that this would be useful for me to know and relevant to the real world (student).
- This is a helpful skill I can use in another subject (student).
- [The task] really got us deeper into the science side of the topic (teacher).

Development of agency or autonomy

- I enjoyed working out what I had to do and when I had to do it (student).
- Students were able to keep trying to create their app until they finally created this amazing new app that they could share with their peers (teacher).

Collaboration and working together

- Students loved working in groups and the conversation was so energetic and inspired (teacher).

Engagement

- The time always goes so quickly when I am in Makerspace (student).
- There is a struggling student who has literacy and numeracy issues but who is so delighted to be in Makerspace it really engages him (teacher).

Reflection

- Feedback was that students were able to demonstrate their deep reflective thinking (teacher).

Excitement

- We love getting the item printed or finished. It is always so exciting, and we can take it home and show our parents (student).
- Students loved having a physical thing like a printed object, something they could use.

1.7 Summary

What is a Makerspace?
A successful, sustainable school Makerspace is grounded in a clear vision of purpose and pedagogy, develops the capacity of the participants (students and teachers), encourages exploration and play, and provides an openness to risk-taking and failure.

Makerspaces can occur within a space that provides resources and opportunities that allow a collective or community to make an artefact/product that is often unique to the maker.

How did Makerspaces develop and evolve?
Makerspaces have been developing organically for many years through Hackspaces and FabLabs. All Makerspaces must have a purpose and meet the needs of all including participants and the school vision.

Teachers must consider the purpose of Makerspace from the moment it is conceived and then approach the practices that best serve its purpose when establishing the space. Teachers need to consider any students' prior knowledge and experiences and then design learning experiences to meet their needs. This could range from explicit instruction to open-ended inquiry, or science focused to including all subjects in STEM, depending on any curriculum requirements. Another possibility includes a more organic and authentic modelling based on what is required.

What are the essential elements of a Makerspace?
Teachers –

- supporting teachers to develop necessary skills and confidence;
- ensuring the support of all the school team including the leadership;
- focusing on students developing hands-on inquiry-based approaches;
- having resilience to overcome failure and setbacks;
- the capacity for teachers to be a lifelong learner;
- preparation to not always be an expert and accept the skills/knowledge brought by students;
- supporting students to effectively work collaboratively;
- ability to look beyond the noise to see the work.

Students –

- encouraging risk-taking to help students overcome failure and push boundaries;
- the capacity to be a lifelong learner;
- the ability to deal with noise and movement in a STEM learning space;
- bringing skills/knowledge to the STEM learning space;
- work collaboratively.

What is Makerspace Quadrant Model and how does it support maker creation?

Makerspace Quadrant Model helps teachers develop the type of Makerspace that best fits their purpose and the pedagogical approach. It also enables teachers to see the wide range of Makerspaces and consider that each has merit in certain circumstances.

Chapter 2
Makerspaces and STEM—A Synergistic Partnership

International research indicates that 75% of the fast-growing occupations require STEM skills and knowledge. (Becker & Parker, 2011)

Keywords Science · Technology · Engineering · Mathematics · Innovation

Definition of STEM

The learning context that determines the emphasis placed on each of the disciplines of Science, Technology, Engineering, and Mathematics and how the various transversal competencies are incorporated into a learning environment.

Focus Questions

- What is STEM?
- Why is STEM important?
- What are the main STEM pedagogies?
- What are the major challenges facing STEM implementation in classrooms/schools?

R. B. Koul et al., *Teaching 21st Century Skills*,
https://doi.org/10.1007/978-981-16-4361-3_2

2.1 Introduction

> To succeed in a new information-based and technologically focused and globalised society, students need to develop their capabilities in STEM beyond what was considered required in the past (National Science Foundation, 2020).

This chapter reviews the origins of Science, Technology, Engineering, and Mathematics (STEM) and considers whether STEM is the answer to the perceived skills crises of the twenty-first century. It examines the history of STEM, and how STEM has been used as a political tool to enable politicians and businesses to financially and ideologically dictate how schools and school systems deliver the 'what' and 'how' of their curriculum.

While not all STEM is taught within a Makerspace environment, this chapter describes how the pedagogy of certain Makerspaces synergizes with the aims of STEM education. These Makerspace environments allow for the development of a number of STEM skills which are usually not evident in the more formalised classroom setting.

2.2 What is STEM?

When academics and teachers talk about STEM, they reach for a definition on which to hang their understanding. The plethora of definitions, however, often creates more confusion than clarity. In laboratories around the world, researchers are solving complex problems bringing together a raft of knowledge and skills in varying amounts and for varying purposes. If we use the 2020 pandemic as an example, we will find researchers combining knowledge from areas identified as Science, Technologies, and Mathematics in order to understand the virus. Regardless of their chosen discipline, their individual and combined aim is to understand the problem and to find a solution. They will rarely consciously or subconsciously articulate 'I am now doing science' but are aware of when they need the expertise of another researcher with different skills and knowledge.

Figure 2.1 shows the cyclical and interrelated nature of STEM subjects. It provides an example of how science and mathematics can explore community needs and provide the necessary content knowledge to ensure these needs are met through solution-designed engineering. The solution produced uses technology, which allows further investigation of the mathematics and science fields. If this was conveyed through vaccine production, it is the science that examines the virus proteins to enable the sequencing and replication of DNA while mathematics considers rates of infection. Engineering is used to synthesise a vaccine from this information, which is multiplied through the use of technology (such as creating cold storage or through delivery) with the efficacy of the vaccine again monitored by science and expressed in mathematics.

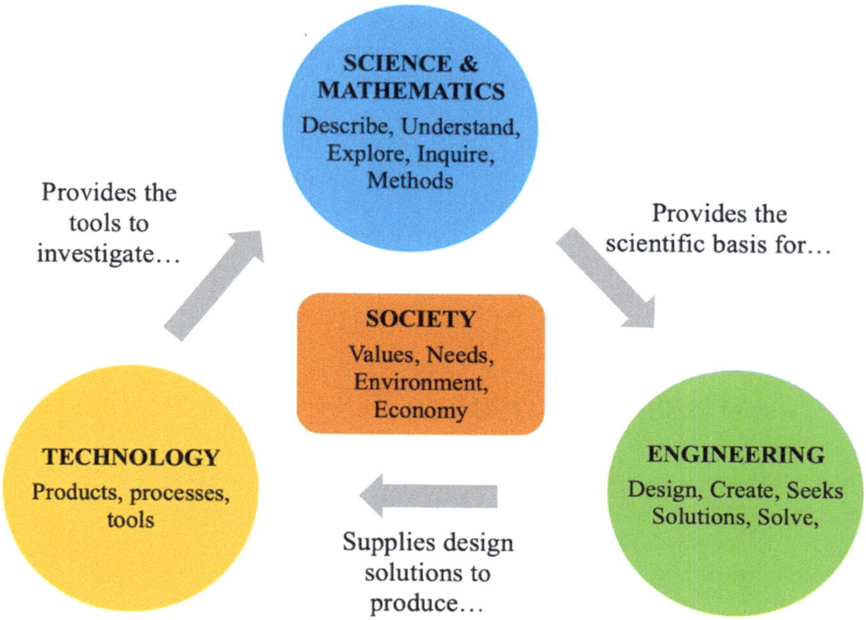

Fig. 2.1 Connectedness of the key areas of STEM (Maynard, 2017)

Advances in STEM have already brought about improvements in many aspects of life, such as health, agriculture, infrastructure, and renewable energy. STEM education is also key for preparing students for the world of work, enabling entry into in-demand STEM careers of tomorrow (United Nations Educational Scientific and Cultural Organisation, 2017; National Science Foundation, 2020). This initially filled a perceived demand from business and industry, which indicated that a population with advanced STEM skills and knowledge would generate solutions to many key issues of the twenty-first century (Blackley & Howell, 2015). Many countries stipulated the need for increased numbers of skilled graduates in STEM that would be equipped to keep abreast of technological advances (Beede, Julian, Langdon, McKittrick, Khan, & Doms, 2011). Across the world, industry experts have determined that creating more graduates in STEM is vital for continued economic success (The World Bank, 2018). The World Economic Forum (2016) also determined that STEM literacy was the measure of the future-readiness of countries. The reported growth in STEM-related jobs was 1.5 times the growth rate of other jobs (14% compared to 9%) between 2006 and 2011 (Timms, Moyle, Weldon, & Mitchel, 2018). In the example of the 2020 pandemic, it is through the modelling of big data (mathematics), the use of technology and engineering to create delivery devices and equipment, and scientific inquiry that has instigated the creation of a vaccine solution to the COVID-19 virus. This issue and the voice of STEM researchers through communication and collaboration shifted worldwide focus towards a cure, which will not come through the fashion, music, or fame careers so feted of late. Consequently, the STEM pipeline—the creation of STEM graduates, developing from primary to secondary school and through to tertiary—continues to be important.

The siloing of learning into subjects in schools has developed out of an evolutionary process of institutionalising learning and education which, in turn, has caused the true nature of learning to be subsumed beneath bureaucratic and political agendas. It appears, however, in recent times that this thinking is now being replaced by a need to return to a more natural, authentic approach to learning with STEM being espoused as the way to do this. STEM is being perceived as a solution to a problem that was created in the education area.

2.3 Global Position of STEM

With United Nations Educational, Scientific and Cultural Organisation (UNESCO), the European Union (EU), and Governments in many countries providing grants for programs to support STEM, it is apparent there is significant financial commitment for programs of this nature. With the rapid increase of scholarly publications on STEM education in recent years, reviews of the status and trends in STEM education research internationally have become increasingly more common. This burgeoning numbers of articles and journals focusing on STEM were highlighted by a recent study by Li (2020) which identified 798 articles in 36 journals from 2000 to 2018. This illustrates how many published empirical studies have been written. In determining the focus on their study, the research team (Li, 2020) reported.

> "a simple Google search with term 'STEM', 'STEM education', or 'STEM education research' all returned more than 450,000,000 items. Such voluminous information shows the rapidly evolving and vibrant field of STEM education and sheds light on the volume of STEM education research."

The study reports that determining how to define the term STEM, as we have already eluded, is complex and confusing and, therefore, makes review and subsequent analysis of STEM complex. Analysis of numbers showed an increase of articles in the last decade with seven articles in 2009 to 230 in 2018 (Li, 2020). It also found that US researchers contributed to 75% of the articles, with Australia (4.7%), Canada (2%), Taiwan, and UK (1.8%), followed by Spain, Israel, South Korea, and Germany (Li, 2020). In Australia,

> "the country and regional reports reveal an almost universal government preoccupation with the level of STEM participation in senior secondary school, and the level of achievement in the STEM-related disciplines in both secondary and higher education. In most nations with active official policy there is also active public discussion." Marginson, Tytler, Freeman & Roberts, 2013).

It was determined that there were a number of reasons for such a high interest in STEM and these related to:

(1) student achievement or lack of achievement in National and International testing such as TIMMS and PISA;
(2) a perceived drop in the numbers of students studying STEM;
(3) a shortage in the labour market of positions in some STEM areas, specifically around new and emerging technologies, and;
(4) a perceived lack of competitiveness in modernisation and technological advantages (Marginson et al., 2013).

As Marginson et al. (2013) suggests, the reasons for the global pre-occupation with STEM subjects, programs, and funding provided for STEM programs are complex and are situated in politics, policy, economics, and culture that make generalised statements unhelpful. Their report, however, articulated that the term 'crisis' was being used in the USA, UK, and Australia; and while there was a relatively small decline in PISA achievements and the numbers of students studying these subjects, the crisis narrative seemed somewhat extreme.

In Australia, like other countries around the world, insufficient school students are studying in Science, Technology, Engineering, and Mathematics prompting increased attention towards student engagement in these subjects. This has consequently led to the creation of 'STEM' education (OECD, 2018). The lack of students is not, however, due to a lack of funding, where the National Innovation and Science Agenda focusing from preschoolers to the tertiary space and into increasing STEM knowledge in the broader community (Australian Government, 2018), has had an investment of $1.1 billion over four years.

In India, there has been a later but significant input of funds into STEM education. Some of these initiatives include:

- *Connected Learning Initiative* supporting 1.5 million students (*CLIx;* https://clix.tiss.edu/);
- A*tal Tinkering Laboratories Project* helping 1 million children (NITI Aayog, 2017);
- *Innovation in Science Pursuit for Inspired Research (INSPIRE) which supports* approximately 200,000 school children every year in the age group of 10–15 years (National Innovation Foundation—India, 2018);
- *National Children's Science Congress* (*NCSC;* http://www.ncsc.co.in/) and the *Initiative for Research and Innovation in Science (IRIS;* http://www.irisnationalfair.org/*)*, and;
- *Science Express (*http://www.sciencexpress.in/) a mobile science exhibition for children mounted on a train which has travelled 142,000 kms and stopped at 455 stations across India.

While there appears to be a large numbers of participants, India has a very large population, so it is crucial to still engage all Indian students in STEM education.

2.4 Defining STEM

The definition of STEM has morphed into many iterations with associated acronyms like STEMM, iSTEM, and STEAM (Sanders, 2009). These are simply STEM with Medicine as STEMM, STEM with Arts to become STEAM, and iSTEM (which means integrated STEM rather than S.T.E.M). The broadest definition of STEM is the study of any of the disciplines—Mathematics, Science, Engineering, and Technology—separately or together. (European Union, 2015; Hackling, Murcia, West, & Anderson, 2014).

The STEM definition on which this book is premised is the partial or total authentic integration of one or more of the STEM discipline areas. There is an acknowledgement; however, that this definition is not comprehensive, and that STEM also needs to include a range of identified skills (Douglas, 2017). Our belief is that STEM should include a range of twenty-first-century skills and competencies which are discussed in Chap. 4 (Transversal Competencies).

Improving the flow of the STEM education pipeline has proven to be a challenge as advocates argue that a more integrated approach is a more authentic view of the world and that adopting this approach would lead to more engaged and interested students (Caprile, Palmen, Sanz, & Dente, 2015; National Research Council, 2011).

STEM education is not simply a new name for the traditional approach to teaching science and mathematics. Nor is it just the grafting of "technology" and "engineering" layers onto standard science and math curricula. Instead, STEM is an approach to teaching that is larger than its constituent parts; it is…a "meta-discipline". STEM education removes traditional barriers erected between the four disciplines, by integrating the four subjects into one cohesive means of teaching and learning (Kennedy & Odell, 2014).

With all the definitions of STEM that are proposed, there is little consideration that in most countries, high schools or secondary schools teach and, more importantly, assess STEM subjects separately (Caprile, Palmen, Sanz, & Dente, 2015; Hackling, Murcia, West, & Anderson, 2014). This has been addressed in some countries, for example in the USA, where engineering education has become incorporated into science education (National Research Council, 2011). In countries such as Australia, Indonesia, and India, however, there is no engineering in the curriculum and subjects in secondary schools are still taught and assessed separately. Currently, it is the assessment that drives the curriculum, and while the subjects are examined separately, no significant changes will be enacted. Despite the intentions of major stakeholders in 'STEM education', there is little evidence of significant systemic change.

2.5 STEM Concepts versus Skills

There seems to be conflict between the narrow views that STEM is focused on the more content-based domains within subjects of Science, Technology, Engineering, and Mathematics and the wider view of STEM in the real world emerging from economic and industry policies. This wider perspective argues that STEM represents more than a collection of content knowledge in Science, Technology, Engineering, and mathematics; that it is 'a point of discontinuity, of society asking for a qualitative change in the objectives of education undertaken in the domain of the sciences' (Lowrie, Leonard & Fitzgerald, 2018).

In more traditional systems (e.g. in the nineteenth century), content knowledge was greatly valued and considered to be of vital importance to graduates and employers. With the advent of the Internet and the escalation in the creation of and access to information, the status of content knowledge has, however, been relegated to the 'back benches' due to its ubiquitous nature. What has now become the true currency of education systems is the far less tangible suite of key employability and professional skills that are today considered, not only essential for the workplace but are part of the focus of STEM education. These include twenty-first-century skills (Partnership 21, 2008), a strong digital focus through the International Society of Technology Education (ISTE, 2016) and the widely encompassing UNESCO transversal competencies (United Nations Educational Scientific and Cultural Organisation, 2015). These represent skills and strategies, not limited to STEM education, which is widely used to measure a potential employee's capacity to be successful in the twenty-first-century workplace. It can be seen here that there is a set of key skills (or competencies) that are consistent in all these frameworks; these include problem-solving, creativity, communication, collaboration, and global citizenship. In this book, these identified skills will be referred to as transversal competencies and they are discussed in detail in Chap. 4.

2.6 STEM Theoretical Framework

When STEM is adopted as a philosophical and pedagogical approach to learning in a school setting, a model or framework needs to be created or adapted and then embedded within the existing curriculum model. It cannot be seen as a separate entity but must align with, or at least complement, the school's existing goals and vision. This is often difficult to achieve when there is only a tokenistic move to include STEM as part of the school program without the belief that what STEM offers can truly transform the teaching/learning paradigm in the school.

The use of a framework or model creates a roadmap that identifies what is important (in terms of what is taught and assessed) and creates consistency in the delivery model. The *Worldly Perspective* (Fig. 2.2) was developed from decades of research in integration by Rennie, Venville, & Wallace (2018) and is not called a framework but a perspective. It shows a balance between the integrated curriculum and the separate areas of Science, Technology, Engineering, and Mathematics and considers the developing perspective from local (me) to global (us). While it provides greater depth, it fails to consider the importance of the transversal competencies/skills that are so prominent in the articulated needs of industry and the wider community.

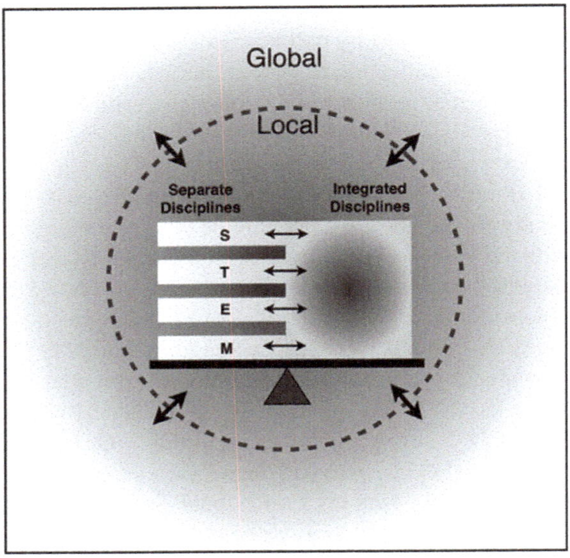

Fig. 2.2 Worldly perspective (Rennie, Venville, Wallace, 2018)

While the Worldly Perspective does consider integration holistically, what is required is a more clearly articulated framework that considers how students develop the key skills or competencies. From our research, we created a model, Model 2.3, that articulates STEM content knowledge in total or partial integration, and the development of transversal competencies. It infuses STEM as a context that provides content that is not measured in the amount of STEM content addressed, but as the context for the depth of the learning for the student (Fig. 2.3).

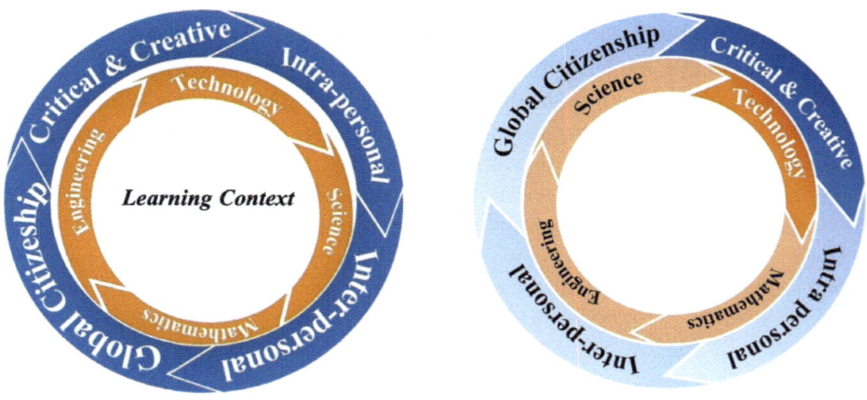

Fig. 2.3 Context-driven STEM model (with example emphasising critical and creative thinking and technology)

2.7 Challenges of STEM

Global challenges to STEM have been identified, and these tensions are seen through curriculum focus, content and structure concerns, teacher-centred vs student-centred pedagogy, and autonomy or accountability.

1. The ***curriculum focus*** we have referred to, with regard to the structure of STE(A)M(M), suggests a wider struggle concerning the skills or generic competencies versus content knowledge through the core disciplines. Teachers are continually under pressure through the curriculum and the subsequent assessments to focus on core content knowledge, and there is often little time made available to teach other learning elements or for a student to take more time to explore a topic more deeply.
2. ***Content versus structure*** in the school relates to the siloed approach to content learning that the majority of secondary schools adhere to. There are content silos such as science and mathematics rather than the structure of the course to focus on inquiry or similar. The siloing of subjects was set up for the support of teachers, and in secondary schools, teachers consider themselves subject experts and not student experts. As the depth of the content increases, it becomes apparent that teachers' depth of knowledge appears only in one subject in secondary school. This does not, however, preclude the idea of school enclaves or communities where teachers can teach across disciplines in a more collaborative mode. In reality, this rarely happens, and teaching can still be a lonely and isolated profession.
3. There is conversation around the type of ***pedagogy*** practised in schools in STEM learning areas. A focus on the value of a student-focused rather than a teacher-centred approach, although not mandated, has been associated with an improved number of students completing STEM in tertiary institutions and then in industry. Teachers who are often required to teach outside their area of expertise are often more afraid to increase students' autonomy as it is a perceived loss of control over student learning which they feel must be under their direction. The time it takes to teach core content knowledge through a teacher-centred approach is usually less than a more inquiry-focused approach, student-focused approach is used. When teachers report being time poor, taking a 'slowly, slowly' approach to the learning is not going to enable them to teach all the perceived necessary content.
4. This also raises the issue of ***autonomy over accountability*** around teachers' knowledge. Teacher accountability concentrates on both the procedures and results of teaching. Both the terms autonomy and accountability are interrelated as when teachers are given autonomy to work in their own way, they must be accountable to consider the consequences. With expensive STEM programs, there is pressure that teachers are financially accountable for the management of the funds and that there are outcomes as well as being accountable for students' learning.

2.8 STEM Pedagogies

There are a number of pedagogies that can be applied to the implementation of a STEM program. We have listed five below.

(1) **Practitioner's Pedagogy Model** This is based on the premise that students come to class with a knowledge base constructed from personal experience(s). From this, knowledge students build a critical reflection of the concepts and ideas enabling them to create and imagine (Luitel, 2020) (Fig. 2.4).

(2) **Transformative STEAM Pedagogies Model** These pedagogies are based on Eastern thinking and focus on the concepts of 'them', 'us', and 'me'. They can be scaffolded as follows:

- Third person (them). Students read or watch texts, websites, and videos that have been written by them or others.
- Second person (us). Students talk to each other and the teacher to make sense of an idea, such as interviews or group discussions (learn from people who are there with them)
- First person (me). Students' own reflections, perceptions, and ideas that are creative and reflective. (Luitel, 2020) (Fig. 2.5).

Fig. 2.4 Practitioner's pedagogy model

Fig. 2.5 Transformative STEAM pedagogies model

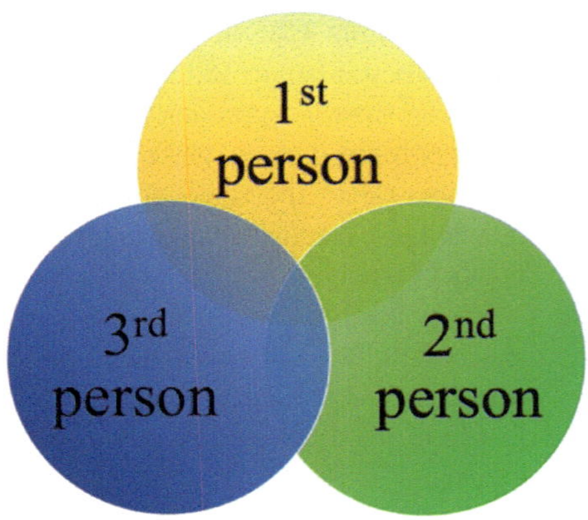

Fig. 2.6 Pedagogical dimension model

(3) **Pedagogical Dimension Model.** This model starts with the (Head) **Cognitive Dimension** at the lower levels including remembering, understanding, and recalling knowledge schema and patterns. The next stage encompasses the **Affective Dimensions** (heart) which includes skills such as empathy to attain higher levels of analysis and creating. **Somatic or kinaesthetic Dimension** (hand) includes body and mind in motion and learning complex ideas which values the practical activities (Luitel, 2020) (Fig. 2.6).

(4) **The Cognition-Based Knowledge Model**. This starts at the lower levels of understanding. For example, what is a Makerspace? This then proceeds through to **Conviction** based which focuses on values, assumptions, and beliefs if there is adequate time to synthesise values and beliefs around a topic. Finally, moving through to **Action-based** knowledge requires us to make changes in our context through actions in the STEM classroom. Before the final action stage, the learner must move through the cognitive and then affective domains (Luitel, 2020) (Fig. 2.7).

(5) **Experiential to Transformative learning**. STEM teaching is seemingly outcomes-driven towards the outcomes specified by assessment and curriculum. If we delve deeply into **Experiential** learning, we can begin to model the experience. For example, if we talk about Makerspaces, then it needs to be modelled for students for their understanding. For STEM learning to become **Transformative**, we must get to a very deep and significant level to transform thinking around an idea or topic. This is a learning environment that is empowering and equitable (Luitel, 2020) (Fig. 2.8).

These pedagogies are all premised by the notion of starting out by finding a simple cognitive stage (such as understanding) before moving deeper into a students' learning. They extend beyond simply recall and remembering by focusing on an active, often hands-on, transformative perspective.

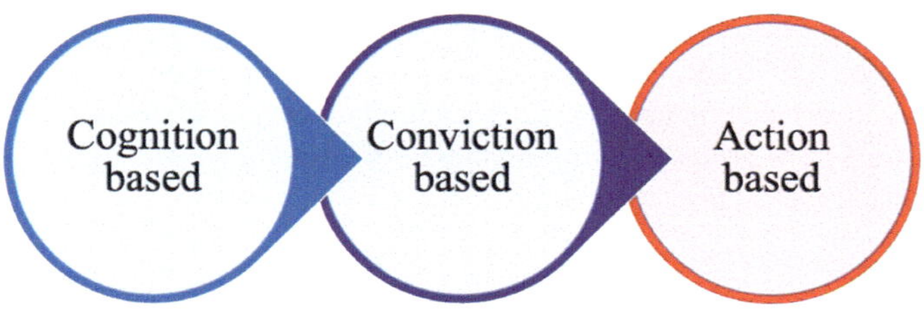

Fig. 2.7 Cognition-based knowledge model

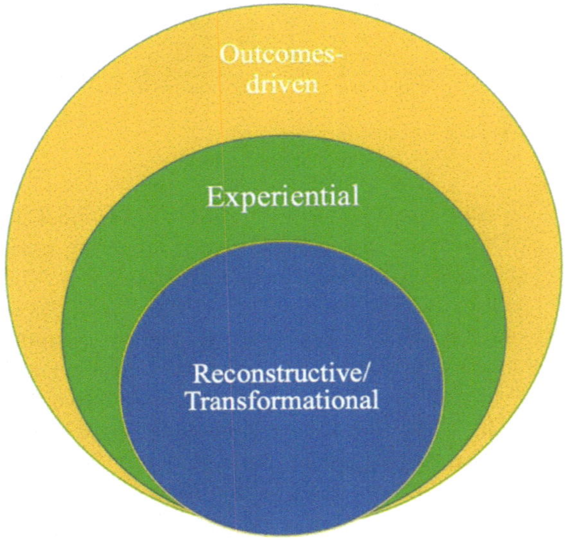

Fig. 2.8 Experiential to transformative learning model

2.9 Makerspace: Bridging the STEM Challenges

Table 2.1 details how Makerspaces are able to address several identified STEM challenges through their innovation, flexibility, and development of twenty-first-century skills in students.

Addressing STEM Challenges through a Makerspace

As discussed in Chap. 1, a Makerspace approach can provide students and educators with a place to refine their skills, knowledge, and play using a 'learning by doing' or experiential approach that has been supported by Papert, Dewey, and Piaget. A Makerspace, if the intentional learning is around STEM as defined in our context-driven STEM model (Fig. 2.3), can include a real local or global problem. Students can be encouraged to ideate, create, and develop solutions or dive into the content knowledge and skills that they will need in order to solve the problem, ultimately developing the skills required in the process. This process can be temporarily scaffolded according to the experiences of the students and the teacher until students gain confidence and competence. While not all STEM is taught in a Makerspace and not all Makerspaces include STEM content, there is a useful synergy that can be seen when the two overlap. It is up to a teacher's imagination how students are engaged with authentic STEM issues by bringing these into classrooms and finding a solution through a Makerspace (Fig. 2.9).

Table 2.1 Resolving STEM issues in a makerspace

Issue	Makerspace
Curriculum focus	Students need to focus on the content knowledge they need at that moment. This can be drawn from a number of content areas (STEM) and can also be skill-based. This enables learning to be messy but rich in multiple curricula
Content or structure concerns	Focus on the breakdown of the timetable where students are able to work on a problem for only the designated lesson time If Makerspaces are used, this time can be extended to be meaningful for the student and Makerspace can exist over multiple curriculum areas, or if Makerspace is outside of the regular timetable, then there can be a multifaceted approach
The teacher-centred versus student-centred pedagogy	The locus of control can be moved towards the students and time can be provided to enable students to move at their own speed to solve the problem that they have identified or is meaningful for them. Makerspaces can be set up with a diminishing scaffold. As the students develop competency in the use of tools and maker-learning, the scaffold can be reduced to give the learner more autonomy

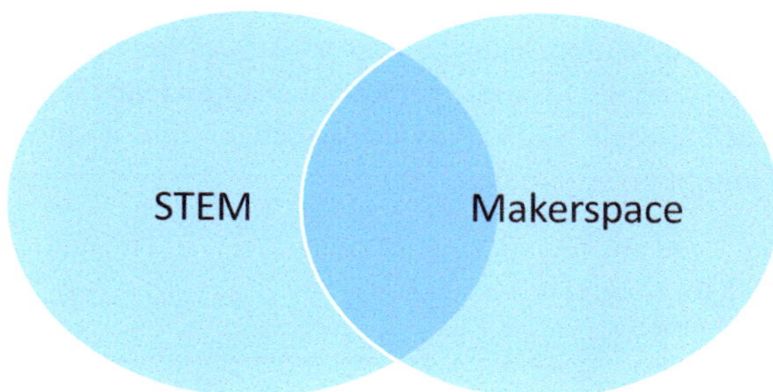

Fig. 2.9 Intersection between STEM and Makerspaces

2.10 Summary

There has been an overemphasis of the power of STEM. In many recent publications, it has been touted as a panacea to solve all the significant current world problems. This promise and the rigidity of many educators and publications around how integration focusing purely on equal contributions of science, technology, engineering, and mathematics has set STEM up to fail. We are proposing that we remove the burden of STEM to be the world's remedy and use STEM Makerspaces as a more organic way of addressing STEM outcomes and encouraging the problem-solving of real-world societal issues. What is also often missing from a STEM approach is the acknowledgement of the importance of key transversal competencies (TVCs), such as critical and creative thinking, communication, collaboration, and resilience. We believe that these essential TVCs build a bridge between STEM subjects by emphasising a natural learning process rather than the artificially siloed learning outcomes in their respective subject areas. Makerspaces provide this environment until education and society understand can understand the crucial intersectionality STEM.

What is STEM?
The learning context that determines the emphasis placed on each of the disciplines of science, technology, engineering, and mathematics and how various transversal competencies are incorporated into a learning environment (Fig. 2.10).

Why is STEM important?
STEM is recognised as the strategy that can build the capacity of a countries workforce to engage successfully in the global economic market.

What are the main STEM pedagogies?
There are a number of STEM pedagogies: these include the practitioners pedagogies, transformative STEAM pedagogies, pedagogical dimensions approach (cognitive, effective, and kinaesthetic), the values-based model, and the transformative model.

What are the major challenges facing STEM?
There are a number of challenges that impede the progress of STEM. These include siloed curriculum and the power of assessments that still drive a content-focused senior school. Disparate pedagogical approaches and a lack of professional development for teachers also provide challenges, as does the lack of autonomy and pressure of autonomy.

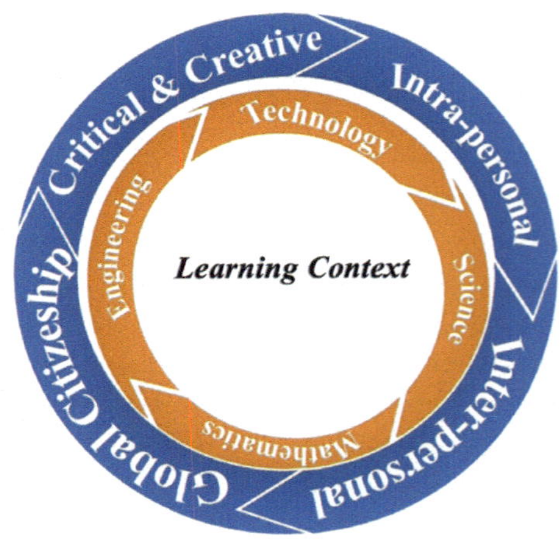

Fig. 2.10 Context-driven STEM model

Chapter 3
The World of Makerspaces

Play is the highest form of research (Einstein)

Keywords Community · School based · Program · Library · Play · Integration · Global

Focus Questions

- Where are Makerspaces found internationally?
- How are library Makerspaces different from those in schools and shopping centres?
- What is ‘purposeful play’?
- How do Makerspaces embrace failure and encourage ‘purposeful play’?

R. B. Koul et al., *Teaching 21st Century Skills*,
https://doi.org/10.1007/978-981-16-4361-3_3

3.1 Introduction

This chapter identifies the purpose, context, audience, and goals of global maker communities and Makerspaces, and the specific strategies employed to ensure their relevance.

Since the early 2010s, makers and making have become fashionable, but as discussed in Chap. 1, the idea of play and constructing is actually much older and taken from the work of Dewey, Wenger, Freire, Papert, and Piaget. We have chosen to use the term 'purposeful play' when discussing Makerspace participation, focusing on constructionism and play with intention. These play areas can be called Hackerspaces, Makerspaces, Maker Fairs, Maker Sheds, Maker Camps, or are named specific to the spaces to which they are connected.

There are thousands of global 'maker' situations, from the presidential Makerspace at the White House, Makerspaces in southern India, playgrounds at the ISTE conference, and Makerspaces in libraries and schools in Australia. There are a vast array of purposes and designs, all of which excite and engage the variety of makers who use them. This chapter reviews the context of Makerspaces (community, school and libraries), their intent and pedagogies, and also considers whether there is a formal or informal STEM focus involved (Fig. 3.1).

Makerspace Quadrant Model (introduced in Chap. 1), Fig. 3.2, is used to position our examples, which include the Malaysia Shopping Centre Makerspace in Kuala Lumpur, Questacon in Canberra Australia, and the Science Centre Makerspace in

Fig. 3.1 President Obama at the first White House Maker Faire in 2014 (US White House)

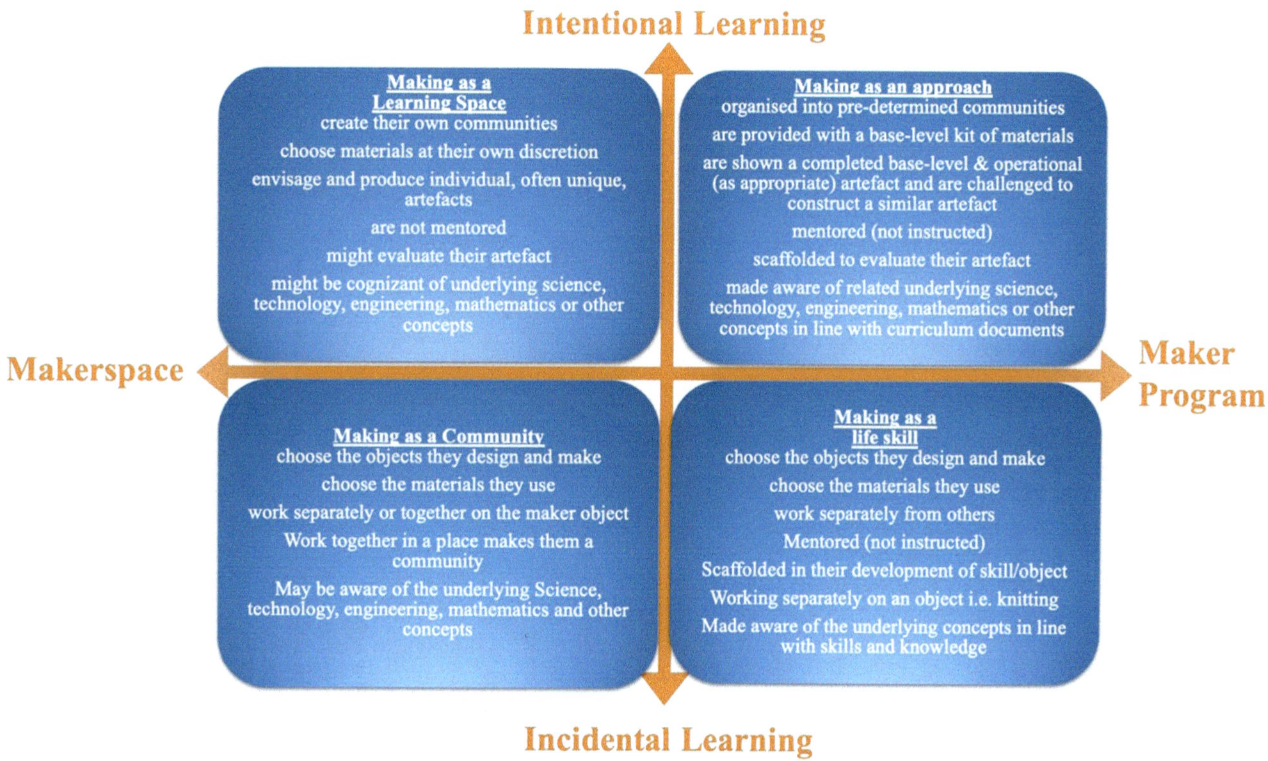

Fig. 3.2 Makerspace Quadrant Model (Vuorikari, Ferrari, Punie (2019)

Indian Institute of Science Education (IISER) Pune, India. Each example is described in terms of whether it provides intentional or incidental learning, and if the focus is on Makerspace or a maker program (Fig. 3.2).

3.2 Makerspace Examples

Community Makerspaces

ME.REKA in Kuala Lumpur in Malaysia

INNOVATIVE DIVERSE COMMUNITY

> Me.reka is an innovative and alternative education space. By removing all barriers to designing and making, we teach students and professionals to excel in the industries and businesses that will shape the future of Malaysia (Me.Reka, 2020).

Named after 'mereka', the Bahasa Melayu word meaning 'to design' as well as a reference to a collective group of people, ME.REKA is a Makerspace based in Publika, Sri Hartamas. This community Makerspace is located in the basement of a large shopping centre near the centre of Kuala Lumpur.

Me.reka describes itself as an organic ecosystem and is designed to provide all the materials, tools, and technologies to support participant's innovation. Members are free to come and use any of the tools and machines that are installed in the building and create any project that they want to, learning the appropriate skills as required.

> The ecosystem at Me.reka brings together students, professionals, communities and makers, and connects them to real projects and income-generating opportunities. Together, we build sustainable solutions and realise our potential (Me.Reka, 2020).

The team at Me.Reka have mapped some of their key opportunities including fabrication and AR against the United Nations Sustainable Development Goals (Table 3.1). Me.Reka uses textiles and other materials and encourages repurposing and recycling to create bags and other items from discarded materials such as seat belts.

Me.Reka has a range of workshop times and large open spaces which enable ***intentional learning*** through specific workshops that focus on teaching concepts such as robotics. These large spaces also provide ***incidental learning*** opportunities where individuals create and develop specific skills that enable them to continue their making journey.

Me.reka currently has two membership plans which can link the members with industry experts. A basic membership allows makers to explore and utilise the space for 20 h per week, access high speed internet as well as a 20% discount on all classes. The pro-membership is aimed at designers and small businesses that have more complex projects in mind which provides 30 h of project time per week, free utilisation of a 3D printer for 1 h, and 5 h on the design computers. Young people, school students, experts, hobbyists, and industry specialists work side by side on their own projects with agency. This can lead to communication, collaboration, and independent learning within the same space.

Table 3.1 How Me.Reka is supporting the United Nations Sustainable Development Goals

TEXTILES Me.reka Textiles Lab encourages users to enquire about work practices and the working conditions of those that make our clothes. Me.reka Textiles Lab focuses on sustainability by engaging users with information on designing circular economy solutions with sustainably sourced textiles and products	
VIRTUAL REALITY Me.Reka Virtual Reality Lab showcases an element of the Fourth Industrial Revolution (IR). Although there are many positives to this 'revolution', millions of people are affected by the disparity provide by digital education, and the lack thereof in certain states and countries. Other concerns include geography, gender, and economy. The onset of virtual reality allows us to be placed in different contexts around the world, and Me.reka's VR Lab allows individuals to educate others on accessible education across the world	

Fig. 3.3 Me.Reka makerspace (Sheffield, 2018)

Teachers are able to bring students to Me.Reka and must be aware of the need for flexibility and aware that they may not be experts themselves. Me.Reka reminds educators that they need to have an open mind and be prepared to be outside their comfort zone. Within Makerspace Quadrant Model, Me.reka is considered as ***a learning space*** and ***a community***. https://www.mereka.my/ (Fig. 3.3).

External School Programs

Questacon Maker Project Canberra, ACT, Australia

SKILLS-BASED SCHOOL PROGRAM

Questacon is an independent organisation that provides a three-storey Makerspace with large exhibit areas where adults and children can come and engage with the space. The organisation also has a second Makerspace located in the Canberra-based *Ian Potter Foundation Technology Learning Centre*, which hosts the Questacon Maker Project. This project offers free innovative workshops for school students to foster an understanding of the process of innovation and manufacturing by supporting hands-on workshops that explore innovation through project-based learning (PBL). The program is delivered in person by the Questacon staff and remotely through interactive virtual excursions delivered via video conference to regional and remote communities. Questacon large open spaces enable ***intentional learning*** through specific workshops that focus on teaching concepts such as problem-based learning and inquiry learning. This is a ***maker program*** rather than a ***community*** approach, and so there is less agency for participants, in this example, students who are brought to the centre to learn specific skills and knowledge (Figs. 3.4 and 3.5).

Fig. 3.4 Ian Potter foundation technology learning hosts the questacon maker project

Fig. 3.5 Ian Potter foundation technology learning hosts the questacon maker project

Makerspace at IISER Pune, India

SCHOOL PROGRAM CONTENT AND SKILLS BASED

Makerspace at IISER Pune consists of a series of activities created by a dedicated Makerspace team that are experts in STEM learning. They have created and demonstrated a variety of STEM-centric activites to undergraduate and secondary students. These experts encourage students to use their STEM understandings to explain the range of phenomena of each activity. This Makerspace is unique in nature, as it focuses on a series of short demonstrations, with makers having limited opportunity to demonstrate agency or decide what to learn and how to learn it. These learning outcomes are highly structured and therefore ***intentional*** and singular in focus (Figs. 3.6, 3.7 and 3.8).

Fig. 3.6 IISER Pune makerspace demonstrations (Sheffield, 2020)

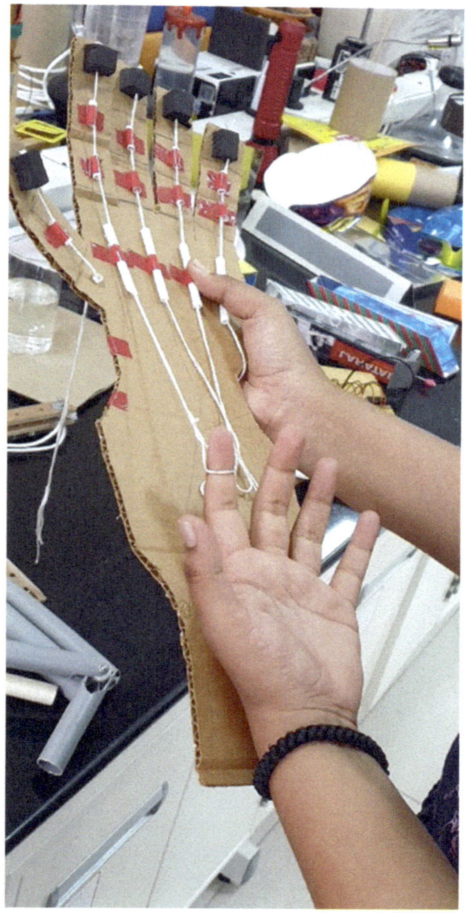

Fig. 3.7 IISER Pune makerspace demonstrations (Sheffield, 2020)

Fig. 3.8 IISER Pune makerspace demonstrations (Sheffield, 2020)

School Makerspaces

Makerspace at the Kindergarten of Wandering Vale Primary School

CLASSROOM TRANSFORMATION INTO A PERPETUAL MAKERSPACE

The teacher in this space consistently applies the maker idea to their classroom space. They believe that a Makerspace should allow children to move around at will and engage with whichever materials suit their desires. Learning is ***incidental*** and includes many opportunities for students to play on their own or with their peers.

This kindergarten had many engaging and exciting maker areas. On the wall shown in Fig. 3.9 shows a wall where students wove various patterns and could climb up on a stool and string up any colour in any design. This woven pattern became an efficient screen to keep the sun out of the play area and provided stimulus for students to start thinking about reflection and shade and shadows (Fig. 3.9).

Figure 3.10 shows how students can work in Makerspace with tools, materials, and other useful objects. Students also had to wear hard hats and gloves to protect themselves (Fig. 3.10).

The materials shown in Fig. 3.11 are also part of the Wandering Vale Makerspace where students can touch, hold, and play with the items, thus providing a valuable kinaesthetic experience. This is very much a ***community*** space for the kindergarten students where the learning is ***unintentional***. Students are able to explore, experiment, and choose what to do and when to do it for the majority of their time in the space. There are some intentional learning opportunities that the teacher had created; however, there was no guarantee that students would choose these activities (Fig. 3.11).

Fig. 3.9 Woven wall as a screen shade at Wandering Vale Primary School (Sheffield, 2017)

Fig. 3.10 Hard hat and hammers at Wandering Vale Primary School (Sheffield, 2017)

Fig. 3.11 Tactile experiences in a Makerspace at Wandering Vale Primary School (Sheffield, 2017)

Makerspace at St Jennifer Catholic Girls School

St Jennifer Catholic Girls School had a purposeful space that was set aside as a permanent Makerspace where students could participate in a range of activities that enabled them to explore.

Designed to support a Junior School STEM (Grades 1–6) and Robotics program, this maker/STEM space was a constantly evolving former classroom learning space, located next to the art room. It was designed to provide a physical and mental space for students to develop computational thinking, problem-solving, and design skills through meaningful engagement with real-world problems and situations. Initially, it was established to support a robotics competition and after-school club, Makerspace

Fig. 3.12 STEM and making: a way of thinking and doing at St Jennifer Catholic School (Graffin, 2019)

became a hub for teaching rich cross-curricular projects in the junior years. It also became the venue for students to meet and share their ideas and projects with real-world experts (Fig. 3.12).

Developing and integrating St Jennifer Catholic Girls School's school-based maker program and physical space required a considerable investment in time and resources; however, the single most valuable investment was in the professional development of the teachers using the space.

St Jennifer Catholic Girls School's maker philosophy promoted STEM and making as *a way of thinking and doing*, encouraging students' collaborative hands-on engagement with real-world technologies and the use of the design thinking or engineering design process to design solutions to real-world problems. This required teachers to learn how to comfortably model and scaffold a flexible, iterative, problem-based approach to teaching and learning. At St Jennifer Catholic Girls School, this was supported by a part-time STEM specialist teacher, who worked collaboratively with classroom teachers to develop and team-teach integrated maker/STEM projects over several years. Through this approach, 'making' became a whole-school responsibility, rather than the domain of a subject specialist (Fig. 3.13).

Fig. 3.13 Micro:bit virtual pet at St Jennifer Catholic School (Graffin, 2019)

Materials and Technologies

As teachers' understanding and appreciation of making evolved over the years, the choice of, and access to, materials and technologies used in Makerspace changed dramatically. Initially, the school invested in expensive 'STEM' gadgets with little understanding of how to effectively integrate these to support teaching and learning. While these gadgets proved to be a fun novelty, they did little to support students' learning or skill development. Over time, teachers and students learned to make extensive use of recycled materials, LEGO bricks and technic, and BBC Micro:bit physical computing devices. These relatively inexpensive tools afforded rich opportunities for integrated cross-curricular projects, including building bug hotels, bushfire emergency warning systems, cardboard automata, and prototyping solutions to problems relating to human spaceflight (Figs. 3.14, 3.15 and 3.16).

Fig. 3.14 Robot dog to help alleviate astronaut isolation on long-duration spaceflight at St Jennifer Catholic School (Graffin, 2019)

Fig. 3.15 Model of a wheelchair friendly cinema, part of a unit on accessible architectural design at St Jennifer Catholic School (Graffin, 2019)

Fig. 3.16 Dancing robot dragons—LEGO robotics with hand-sewn costumes at St Jennifer Catholic School (Graffin, 2019)

Fig. 3.17 LEGO WeDo-based robotic vehicle for underground mining at St Jennifer Catholic School (Graffin, 2019)

An inclusive LEGO Robotics program Traditionally, LEGO robotics and robotics competitions have tended to focus on the learning needs of gifted and talented students. This is primarily due to the expense of the equipment and the time teachers need to learn how to scaffold robotics projects. St Jennifer Catholic Girls Junior School took an alternative approach, initially starting with a small, but highly inclusive LEGO robotics competition program open to any students with an interest and aptitude for visual programming and engineering, regardless of their academic standing. These students have since competed in, and won, state and national competitions. Drawing upon the lessons learned during this extraordinarily successful competition program, the school recently started integrating LEGO robotics across the curriculum and year levels, especially in early childhood (Fig. 3.17).

Makerspace at Elmtree Primary School

INTENTIONAL LEARNING MAKERPROGRAM FACE TO FACE AND ONLINE CREATING

Elmtree Primary School uses a blended pedagogy approach comprising the Walker Learning approach K-2 and an inquiry-based approach in Years 3–6. Their focus on STEM, which is integrated within these learning approaches, specifically seeks to engage communities of practise external to the classroom and apply them to classroom learning. The advantageous nature of forming relationships with experts beyond the school allows children to develop real-world connections and be inspired by individuals within a range of contexts.

At Elmtree Primary School, Miss Claire is a driver of the development and professional learning of STEM and inquiry education. Her Year 5 classroom Makerspace was developed to facilitate ***intentional learning*** and explicit teaching of competencies, driven through student autonomy and choice. She adopted what we have defined as a ***maker approach*** rather than a Makerspace. It was designed for specific learning outcomes, having both the knowledge and cognitive domains in mind as it was created. Students are immersed within contextual experiences that are designed to explicitly teach both skills and knowledge, which position students to become independent and confident problem-solvers over time (Figs. 3.18).

Fig. 3.18 Cooperative-based learning environment to support autonomy and choice at Elmtree Primary School (Fairhurst, 2020)

Fig. 3.19 Miss Claire's physical Makespace room where students are supported flexibly with resources, tools, and inspiration at Elmtree Primary School (Fairhurst, 2020)

Miss Claire's learning environment is designed to foster curiosity and to support students to extend themselves through explicitly driven inquiry, personal passion projects and cooperative teamwork. Students are given access to a wide range of resources, tools, and options to design and develop solutions to problems set within contexts of the inquiry. This is inclusive of physical and online Makerspaces. An example of this was during a sustainability inquiry about Western Australian bees. Miss Claire worked with her students to develop both a space where students transition between a physical Makerspace and a virtual Makerspace within the Minecraft realm (Fig. 3.19).

Fig. 3.20 Examples of a physical student learning routine at Elmtree Primary School (Fairhurst, 2020)

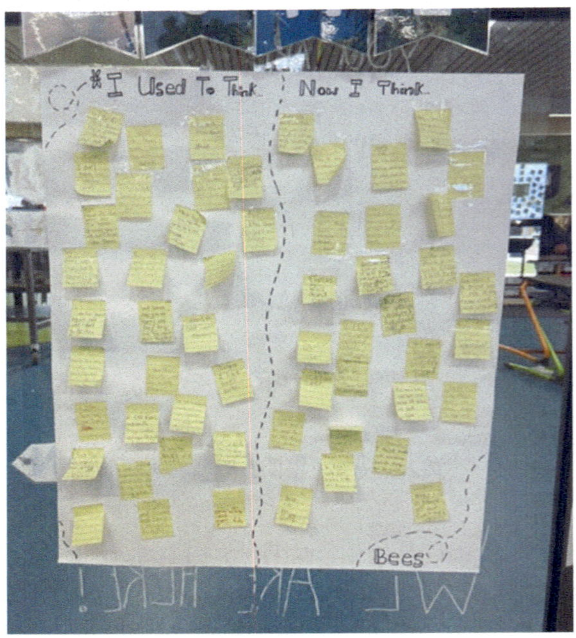

Fig. 3.21 Examples of a physical student learning routine at Elmtree Primary School (Fairhurst, 2020)

Initially, students interacted with a community-based expert through Webex, where they were hooked into the project through real-world applications. The students were then challenged to develop a research knowledge base about Western Australian bees through the use of cooperative teamwork and Visible Thinking Routines (Harvard University). Miss Claire's class then reported back their findings to their expert over Webex, and through this interaction, her students were further challenged to independently determine a means of communicating their findings to impact people beyond their classroom (Figs. 3.20 and 3.21).

Virtual Makerspaces

Within Miss Claire's intentional Makerspace, student choice and autonomy are celebrated. The students determined a range of ways to communicate their findings about their bee project and were supported to achieve these successfully through their learning environment.

Her class decided that they wanted to create a virtual environment to engage as many people as possible with their learning, so Miss Claire organised a virtual Minecraft field trip for the children to undertake so that they could experience and

Fig. 3.22 Student-developed Minecraft Education behaviour expectations at Elmtree Primary School (Fairhurst, 2020)

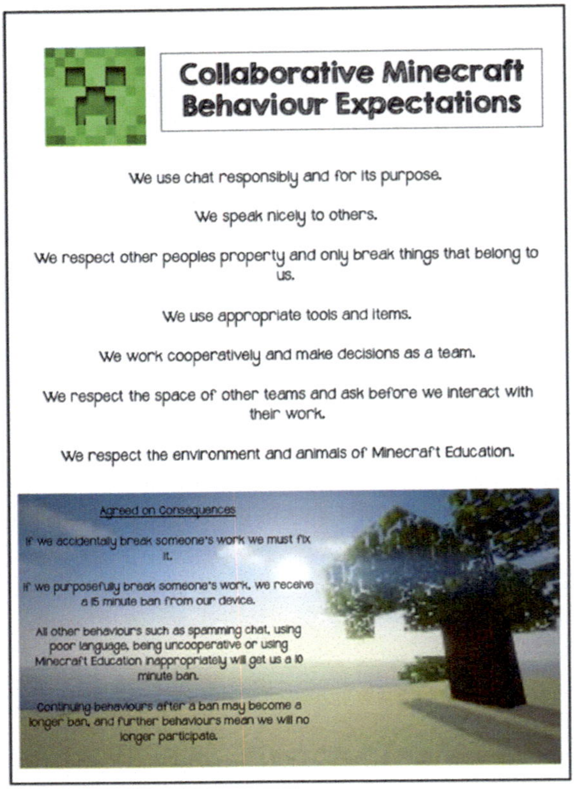

Fig. 3.23 Student-developed Minecraft Education behaviour expectations at Elmtree Primary School (Fairhurst, 2020)

research the type of space they wished to develop. In order to achieve this, she spent time with her students creating a set of expectations for virtual excursions—something the students had not experienced prior. Within their teams, the children brainstormed the expected behaviours they wanted to see from each other, and then as a class they came together to create the document. Miss Claire strongly believes that students must be responsible for writing the behaviour expectations, and that they are revisited frequently as students experience frustrations and conflict with one another. She also firmly believes that they must choose the consequences for their behaviours, allowing them to be in control of their social contract with each other (Figs. 3.22 and 3.23).

Having determined their behaviour expectations and having discussed potential consequences for not displaying these behaviours, Miss Claire hosted the Minecraft Education biological sciences map about American bees. The teacher and her students travelled together through the world looking at how this map had been created to teach others about bees within America. They discussed the use of interactive stations, Non-Player Characters (NPCs), signposts, and aesthetics to engage people with their learning. From this, success criteria were created by the students about the features

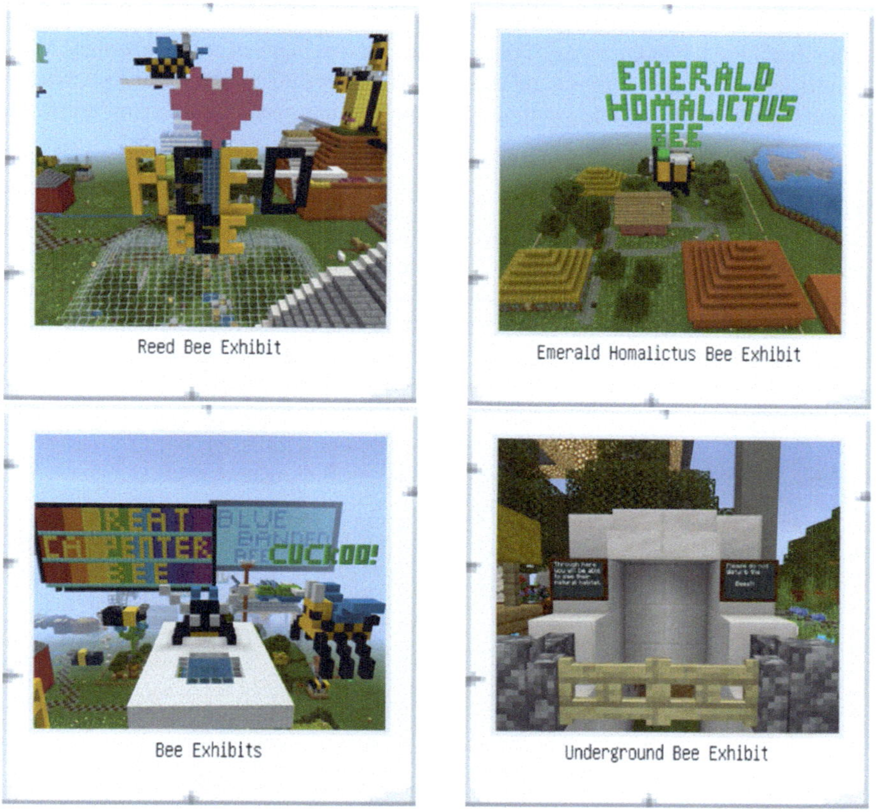

Fig. 3.24 Examples of student collaborative work within the Western Australian Bee map at Elmtree Primary School (Fairhurst, 2020)

they wanted to include within their Minecraft Australian bee map. The students then formed teams based on the type of bee they had researched. Miss Claire's class already had Minecraft community behaviour expectations the students had collaboratively created from previous projects; however, these expectations were revisited and agreed upon with the students prior to them joining the community map.

Miss Claire then gave the students the freedom to begin developing their areas within their teams and supported this development through the setting of parameters and giving constructive feedback to builds. Explicit goals were also given in terms of their success criteria, specifically around the amount of information required about the bees and the interactivity of the bee exhibits. Over several sessions, the students worked on the collaborative map to create their Australian Bee Park to share with their community. One of the most interesting responses from the students based on their learning was their collective desire to contact Minecraft Education about their misrepresentation of bees. The only way that students were able to spawn bees for their map, was to create hives that would fill with honey. They felt confused by this, stating that there are many types of bees that do not live in hives or create honey. Miss Claire discussed with her class and they decided to create exhibits that represented this and showed the bees living in their actual natural habitats, such as in underground burrows (Fig. 3.24).

Physical Makespace—Bee Hotels

Along with the communication of learning through the virtual Makerspace, Miss Claire also worked with her students to purposefully and flexibly create a physical Makerspace that represented their learning.

Responding to student interest in bee habitats, and utilising a long-standing partnership with Curtin University, Miss Claire organised an incursion for students to design and create their own 'bee hotels'. Students had proposed that one of their

Fig. 3.25 Student wearing safety glasses and using a hacksaw to cut through PVC pipe at Elmtree Primary School (Fairhurst, 2020)

Fig. 3.26 Examples of student bee hotels at Elmtree Primary School (Fairhurst, 2020)

Fig. 3.27 Examples of student bee hotels at Elmtree Primary School (Fairhurst, 2020)

Fig. 3.28 Examples of student bee hotels at Elmtree Primary School (Fairhurst, 2020)

methods of communication would be to create a 'bee garden' within the school, and one way of starting this project was to create homes that would attract native bees. Students worked in teams to develop their bee hotels and were mentored by adults to use a wide range of tools, such as hacksaws and drills, to create their projects. These skills are highly valued in Miss Claire's classroom, with the idea that students can learn to use potentially dangerous tools in a safe and scaffolded environment, developing essential life skills (Fig. 3.25).

Students had open choice of resourcing for their project, allowing them to test materials that they thought would be weather-proof and strong enough to withstand the strong winds of the Mandurah region. In their terms, they negotiated their designs, basing their choices on their bee research.

They problem-solved collaboratively and were supported in their decision making when necessary. Conversations involved discussing types of materials required for different types of bees, the positioning of the materials, and reflecting upon potential mistakes during the building process and how to fix them (Figs. 3.26, 3.27 and 3.28).

Bee Art

Another way the students proposed to communicate was through a physical art exhibition. The students decided they wanted to contact an artist to help them create ways to teach their school community through a medium they had been learning within their specialist art classes. Miss Claire contacted a local Mandurah artist who was interested in working on the project with the class. She helped to bring the students ideas to life with her specific skill sets and assisted with navigating the project.

The students decided to create a 'bee garden' exhibition, where they would use recycled fabrics and textiles to create representations of the bees that they had studied. These would be placed around the bee hotels that the students had created as part of their physical STEM project with Curtin University. One of the key ideas that also

Fig. 3.29 Artistic representations of the blue banded bee, emerald homalictus, and the great carpenter bee at Elmtree Primary School (Fairhurst, 2020)

Fig. 3.30 Artistic representations of the blue banded bee, emerald homalictus, and the great carpenter bee at Elmtree Primary School (Fairhurst, 2020)

Fig. 3.31 Artistic representations of the blue banded bee, emerald homalictus, and the great carpenter bee at Elmtree Primary School (Fairhurst, 2020)

Fig. 3.32 Collaborative art piece with a local artist representing the blue banded bee at Elmtree Primary School (Fairhurst, 2020)

came from the bee garden project was around the types of flowers that we would want to plant to attract the bees. Students created artistic representations of flowers such as lavender and other bee-friendly plants that they included within the art display (Figs. 3.29, 3.30, 3.31 and 3.32).

The Mandurah artist also supported the students to create a collaborative art piece about the blue banded bee, a favourite of the classroom due to its unique colours. Students took turns to assist in the creation of the artwork which is on display at the school.

Student Interview Responses

Due to the innovative nature of her physical and virtual learning environment, Miss Claire conducted an interview with her class to capture their feelings about physical and virtual learning environments, including excursions and incursions. Overall, students had highly positive feelings about both learning environments. Some interesting comments made by students were that sometimes they felt anxiety about leaving the school on a physical excursion, and that virtual excursions were a lot more relaxed. They highly valued the hands-on and interactive elements of physical incursions and excursions and felt more creative during these times. Some students commented on how a change of setting or routine could sometimes be frightening, and that the noise of the classroom is usually higher than usual.

The students also commented on the structure of physical environments during the interview, and made comments about how easy it is to 'mess things up' when there is less structure. However, they noted that they knew that making mistakes was okay and that they knew there were supports in place to help them if they needed it. They also noted that freedom could be a good and bad thing, stating that sometimes they do not make the best choices in their teams and that they need to be more in control of their own behaviour.

During the interview about virtual learning environments, students noted that they felt safer and could easily work collaboratively with their peers. Some students commented on resourcing, explaining that in a virtual world such as Minecraft, they had greater freedom of choice with their presentations and had a wider range of resources as they are free and infinite on a virtual platform. An interesting comment made by a student was that 'teachers may not feel that it is the most structured thing for school, but students will most likely do work the best they can on technology cause we love it'. Miss Claire's students also commented on the way that teachers can easily behaviour manage within virtual Makerspace environments, explaining that they were able to use their behaviour expectations to report behaviour as necessary. They also commented on the functionality of the chatroom space within the virtual Makerspace, explaining that they could easily use this to communicate and help each other learn new skills or develop their projects further.

A disadvantage noted by the students was that they had more opportunities for destructive behaviour on virtual Makerspaces, including setting fire to other student projects. However, they noted that through their cooperative teams they would be able to respond quickly as a group to solve the problem. Another disadvantage noted by students was also about resourcing, but explaining that an infinite amount of resources could also mean an endless project where students find it hard to decide when they have finished.

3.3 Makerspaces in Libraries

Following the 2014 White House Maker Faire, over 100 ULC member libraries signed a letter to President Obama espousing the virtues of libraries as an incubator for Makerspace innovation.

> **Dear President Obama:**
> The mission of libraries is to improve society through facilitating knowledge creation locally and globally. Library Makerspaces mobilize communities around STEM learning initiatives, foster a spirit of innovation and entrepreneurship, and create universal access to ideas and information. Makerspaces benefit libraries by expanding the educational resources available to citizens, reinvigorating the library experience, and increasing community

engagement. Library Makerspaces benefit citizens by allowing them hands-on engagement opportunities, opening access to modern software and hardware resources that allow citizens to design, create, and make almost anything they can dream up… Library Makerspaces bring people together to collaborate, to remix, to participate in the creation and recreation of their culture… Makerspaces in libraries can contribute to this effort, allowing everyone to develop critical thinking and problem-solving skills and facilitate opportunity for collaboration and community engagement that will aid in the next generation of STEM jobs. They provide access to tools (from books to 3D printers) and, most importantly, 'access to each other'. Library Makerspaces are powerful informal learning spaces that give local community members the ability to design, create, and innovate their future.

Libraries have long been a hub of special activities within the school or community. Key events (e.g. Book Week), special days (Australian Day and ANZAC Day), guest speakers and authors, performances, hands-on activities, and much more infuse the library program throughout the year. 'Make and take' arts and craft activities are common library activities with students. Therefore, the transformation into a more targeted STEM or STEAM focused program is but a small stepping-stone from these traditional activities into a more targeted Makerspace. Student curiosity is encouraged, nurtured, and rewarded as their learning is personalised, and their interests are supported. Any student can come to the library to get the help and resources they need to be successful. In the age of print, a library was perceived as a repository of information and knowledge. With the digital revolution computers, online databases, printers and eBooks became the norm (Wang, 2016). Now, add to this interactive, digital learning technologies for content creation such as filmmaking and editing (including green screens), recording studios, programming/coding, robotics and 3D printing, craft making equipment and materials, and what you have is space where students have new opportunities to collaborate, problem solve, build, investigate, and produce in a non-threatening way (Britton, 2012)—the perfect foundation on which to build a Makerspace.

Makerspaces foster all the best things the library has to offer:

- A space to satisfy and discover curiosities;
- An opportunity to engage in self-directed learning based on independent interests;
- An opportunity to connect with adult and peer mentors around a given interest; and
- The chance to create and share those creations in a community venue that celebrates discovery, their capacities and accomplishments.

Bowler and Champagne (2016) believe that well-designed Makerspaces in libraries can enable a 'shift in users experience from one of passive consumption to another of active production'.

Librarians have always been advocates of twenty-first-century skills mainly through their inquiry and problem-based learning programs; therefore, the adoption of student-centred and open-ended activities into a makers program is a manageable leap for most information specialists. This transformation, however, needs to be undertaken through a rigorous and thorough consideration of the intended outcomes. While the purpose of the library Makerspace needs to be considered within the mission statement of the school library, and align with its goals and objectives, it is important that library staff understand and are able to articulate the reasons why their library provides a Makerspace for students. For example, some libraries have elected to focus solely on a specific area, skill set or equipment (3D/Laser Printing), while others focus on coding, electronics or digital art (e.g. CoderDojo or Hour of Code). Regardless of the format Makerspace takes, there are some basic questions that need to be considered long before the library Makerspace is launched:

- Why should the library have a Makerspace?
- What is the library trying to achieve via its Makerspace?
- How does a Makerspace fit into the library's mission/vision?
- Who are our target clients and are their unique needs being addressed through access to a Makerspace?
- Where should we start (Velasquez, 2018)?

Once these questions can be answered and the purpose and structure of Makerspace is identified, the transformation or enhancement of the library programs and services can occur. The following example illustrates how a tertiary library has taken the initiative to champion the creation and ongoing management of a Makerspace at a University Campus, thus showing successful transformation from traditional library service to facilitator of a futuristic vision of learning.

3.4 Makerspace at Curtin University, Western Australia

SKILLS-BASED TERTIARY COMMUNITY PROGRAM AND SPACE

Like Re.Meka, the Curtin Makerspace, has a range of **Maker community** opportunities where makers can use the space to create the projects that they want to when they want to. There are also **Maker programs** where the makers learn a specific skill under guidance at a specific time. Learning can be **intentional or incidental depending** on the activities or experiences undertaken.

Curtin Library started with a small room that was very limited in materials on the ground floor of the building. The Arduino, Raspberry Pi, and other electronics were not able to be borrowed and it was not supervised, so access time was limited. After a small community grant during science week, the library moved Makerspace to a much larger space on the 5th floor. It describes itself as a physical 'hub' or learning space for the community of makers within, or connected to, the Curtin community. We also seek to engage online, and with the wider community to build relationships

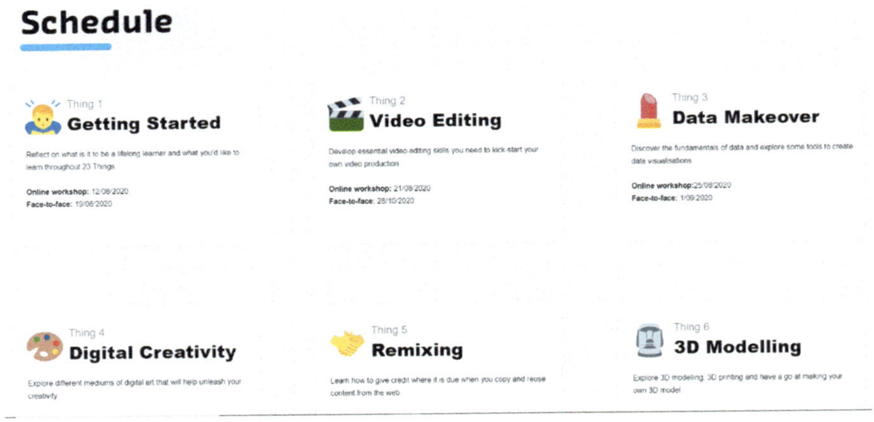

Fig. 3.33 Curtin Makerspace program and schedule

and collaborative partnerships. In so doing, we aim to strengthen the reputation of the Curtin Library, and that of Curtin University, as a driver of innovation (Curtin University, 2020). There are a range of workshops scheduled under areas or you are invited to knock on the door, and they will try and find room for you. Joining the live or online community, Makerspace is open to Curtin University members including staff, students and alumni (Fig. 3.33).

3.5 Summary

To conclude this chapter, we have taken Makerspace Quadrant Model elements discussed at the beginning of the chapter and identified where our examples 'sit' within the model (Table 3.2).

Where are Makerspaces found internationally?
Makerspaces can be found in a variety of locations including schools, shopping malls, universities, and community centres. On a global scale, Makerspaces are present in Australia, Malaysia, India, and the USA, among others.

How are library Makerspaces different from those in schools and shopping centres?
Each Makerspace needs to be fit for purpose and ensure that it meets the safety needs and the needs of the makers within the space. Any further generalisations are unhelpful as Makerspaces can be transient, permanent, all-encompassing or narrow and real or virtual.

Table 3.2 Summary of the international Makerspaces examples and where they ft into Makerspace Quadrant Model

	Learning intent		Pedagogy		STEM	
	Intentional	Incidental	Space	Program	Formal	Informal
Community						
Me. Reka in Malaysia		X	X			X
Questacon Australia	X			X	X	
IISER Pune	X			X	X	
School						
Wandering Vale		X	X			X
St Jennifer's	X			X	X	
Elmtree Primary School	X			X	X	
Library						
Curtin University Library		X	X	X		

What is 'purposeful play'

We refer to purposeful play and the work of Dewey, Wenger, Freire, Papert, and Piaget when discussing Makerspace participation, focusing on constructionism and play with intention. These play areas can be called Hackerspaces, Makerspaces, Maker Fairs, Maker Sheds or even Maker Camps.

How do Makerspaces embrace failure and encourage 'purposeful play'?

Makerspaces are mostly low risk spaces where students are not distressed if they 'fail' in their initial attempt. If Makerspaces become more mainstream and they form part of the learning and assessment of students' work, then they become more high risk and students are more likely to follow the perceived rules and not 'play'.

Part II
Makerspaces and Pedagogy—Learning in a Makerspace

Preface
Part II explores how twenty-first-century skills can be developed within a Makerspace context. These skills, often referred to as transversal competencies, are mapped in Chapter 4 against international frameworks to highlight their importance. Subsequent chapters in this section identify five twenty-first-century skills that can be readily developed within a Makerspace.

Chapter 4. "The Essential Twenty-first-entury Skill Set—Transversal Competencies"
This chapter explores how the twenty-first-century skills, general capabilities, and transversal competencies align and create a framework of the key definitions for these terms. It examines the development of these 'soft skills' and considers their current prominence across the world. The twenty-first-century skills in the European Union are examined and contrasted against the UNESCO's transversal competencies and the Australian General Capabilities.

Chapter 5. "Developing Higher-Order Thinking in a Makerspace"
This chapter defines higher-order thinking and provides guidelines on how to nurture critical and creative thinking within a STEM Makerspace. Bloom's Taxonomy is used to show the importance of effective questioning to develop higher-order thinking, and strategies are provided to help educators promote thinking in the classroom. A case study illustrates what creative thinking can look like within a Makerspace setting.

Chapter 6. "Developing Collaboration Skills in a Makerspace"
This chapter discusses collaboration as a dynamic process and describes how educators can develop a collaborative STEM Makerspace culture. The Build, Act, Review (BAR) model is used to explain the pedagogical approaches to develop essential collaboration skills. The case study highlights how student participation in hands-on activities can promote collaborative skills and how these skills can be measured by the classroom teacher.

Chapter 7. "Developing Communication Skills in a Makerspace"
This chapter defines the characteristics of communication and describes how a STEM Makerspace can develop key communication skills. The BAR model is used as a framework for developing effective questions by both the teacher and students. The

case study highlights how student participation in hands-on activities can promote communication skills and how these skills can be measured by the classroom teacher.

Chapter 8. "Developing Resilience in a Makerspace"
This chapter defines the key skill resilience. A number of reflective strategies are described to promote and map resilient behaviours in students. This chapter defines the trait of resilience and describes reflective strategies that can be used to promote and map resilient behaviour in students. The case study highlights a student's struggle to develop resilience in Makerspace and the importance of reframing failure.

While we have identified these competences in four distinct chapters, we acknowledge the fact that they are indeed interconnected or 'entangled'. When interrogating each competency, there is an overlap. For example, students were often demonstrating communication and collaboration skills at the same time or collaborating with others required a certain level of resilience. The fact that the four C's (communication, collaboration, critical, and creative thinking) are regularly grouped together also highlights the interconnectedness of each of these competencies.

Chapter 4
The Essential Twenty-First-Century Skill Set—Transversal Competencies

Skills have become the global currency of twenty-first century economies. (UNESCO, 2015)

Keywords Soft skills · Twenty-first-century skills · General Capabilities · Transversal Competencies

Transversal Competency Definition
A broad-based set of skills, knowledge, and understandings considered essential to live and work in the twenty-first century. They are also sometimes referred to as twenty-first-century, soft, or entrepreneurial skills.

Focus Questions

- What are Transversal Competencies?
- Why are they important to develop in students?
- What do these skills look like?
- How can these skills be taught in a Makerspace?

R. B. Koul et al., *Teaching 21st Century Skills*,
https://doi.org/10.1007/978-981-16-4361-3_4

4.1 Introduction

> The globalisation and internationalisation of the economy along with the rapid development of information and communication technologies (ICT) are continuously transforming the way in which we live, work, and learn." As the world shifts from a resource-based to a knowledge-based economy, the impact on the types of jobs available and the work skills needed to secure employment are both increasing.
>
> Voogt (2012)

In 2006, the European Parliament and Council adopted the *Key Competencies for Lifelong Learning* which were defined as the competencies each European citizen needs for personal fulfilment and development, employment, social inclusion, and active citizenship (European Commission, 2018).

Then, in April 2018, the European Parliament and Council (EU) 2018/646 acted again to formalise this process further by articulating a common framework for the provision of better services for skills and qualifications (Europass). They suggested that 'transversal or soft skills, such as critical thinking, team work, problem-solving and creativity, digital or language skills, are increasingly important and are essential prerequisites for personal and professional fulfilment and can be applied in different fields' (European Commission 2018).

Since these two major initiatives, much has been written about the new skill set required of individuals to effectively work and live in the current economy and, what originated in Europe, has now become a global imperative. According to McKinsey (2017), workers of the future will spend more time on activities that machines are less capable of, such as managing people and communicating with others with less time spent on predictable physical activities and data collection and processing. The skills and abilities required will also shift to more social and emotional skills with a greater need for more advanced cognitive abilities such as logical reasoning and creativity.

The heightened recognition for this hybrid skill set to become an integral part of the curriculum has placed increased pressure on educational institutions to respond accordingly (Redecker et al., 2011). While these skills are not new, their importance as part of the formal education process has shifted significantly and now their assumed adoption and ad hoc implementation has become a structured focus for learning programs. As curricula evolve and imbue new terminologies that reflect these required skill sets, educators must engage with and operationalise them in the programs they teach.

Perhaps the most significant attempt to make sense of what this essential skillset might look like in an educational context came with UNESCO's (© United Nations Educational, Scientific and Cultural Organisation) *Transversal Competencies in Education Policy and Practice Report* (2015). This document suggests that education policies and curricula must aim to incorporate a broad range of skills and competencies for learners to successfully navigate the changing global landscape. This report also emphasises that curriculum needs to ensure that students develop attributes and skills necessary for a rapidly changing society and workplace (© UNESCO, 2015).

Gonski (2018) further supported and gave momentum to this premise in Sect. 2.5 of the *Report of the Review to Achieve Educational Excellence in Australian Schools* by stating 'every student needs to be equipped with the skills and knowledge to navigate a rapidly changing world' (2018).

4.2 What is in a Name?

Various terminologies have been explored by educational institutions, international organisations, and within the research community to capture, compartmentalise, and name the shifting cluster of competencies required in twenty-first-century learning as there is no definitive way to describe them collectively (Gonski 2018). Terms in use include 'twenty-first-century skills' or 'twenty-first-century learning' (used by *The Assessment and Teaching of 21st Century Skills* [ATC21S] and *Partnership for twenty-first century skills* [P21], 'key competencies' (OECD 2005), 'soft skills', 'STEM skills', 'new collar skills' (McKinsey 2017), and 'entrepreneurial skills' (New Work Smarts, 2017). While the term 'twenty-first-century skills' is widely used, many educators argue that the skills and capabilities referred to were important well before the twenty-first century, while also noting that with this rapid change, century-long milestones are inappropriate (Voogt 2015).

One term that is gaining wide acceptance is Transversal Competencies (TVCs) which has re-emerged as a way of describing these broad-based skills, knowledge, and understandings. The European Commission explicitly used this term when they suggested that:

> "many national curricula have moved towards integrating transversal competencies as a response to the number of social, economic, and cultural changes brought on by globalisation and the rapid development of Information and Communications Technologies" (Voogt 2002).

UNESCO also uses the term Transversal Competencies to describe skills that aim to meet current and future challenges (© UNESCO, 2015).

4.3 Developing Competency Frameworks

A consequence of attempting to label these skills and competencies is the development of numerous frameworks that schematically describe the relationships between and hierarchy of essential workplace and educational skills. In Voogt's 2012 study 'A comparative analysis of international frameworks for twenty-first-century competences', he found they were largely consistent with their inability to define twenty-first-century competencies; however, each framework had a different focus and area of emphasis within the overarching competencies (Huiying, 2017).

Table 4.1 describes the ATS2020 (2018) categorisation of several frameworks that apply to an educational setting.

Table 4.1 Description of digital competences / skills frameworks

Digital Competences / Skills Frameworks	General Frameworks of Twenty-first-century Skills	National Key Skills Frameworks
UNESCO ICT Competency Framework for Teachers (2011)- an 18-module framework that focus on teaching and learning in 3 domains: Technology literacy, Knowledge deepening and knowledge creation (http://unesdoc.unesco.org/images/0021/002134/213475e.pdf) (Amzat, 2016)	***P21 Framework for 21st Century Learning (2015)*** Content knowledge and twenty-first-century themes, Learning and innovation skills Information, media and technology skills and Life and career skills (http://www.p21.org/our-work/p21-framework)	***ISTE NETS (2015)*** - support students, educators and leaders with clear guidelines for the skills, knowledge and approaches they need to succeed in the digital age (http://www.iste.org/standards/iste-standards)
General Capabilities in the Australian Curriculum Literacy, Numeracy, ICT capability, Critical and creative thinking, Personal and social capability, Intercultural understanding and Ethical understanding. (http://www.australiancurriculum.edu.au)	***21CLD - (twenty-first-century Learning Design) (2013)*** - six skill domains collaboration, Knowledge construction, Skilled communication, Real-world problem-solving and innovation, Use of ICT for learning and Self -regulation. (https://www.sri.com/work/projects/21st-century-learning-design-21cld) (Poysa-Tarhonen, 2018)	***ATC21S (Assessment and teaching of twenty-first-century skills) (2012)*** - defined ten twenty-first-century skills into four broad categories: Ways of thinking Ways of working Tools for working Living in the world (http://www.atc21s.org/) (Poysa-Tarhonen, 2018)

Consequently, the importance assigned to a specific set of competences varies from framework to framework (Voogt, 2012). While some frameworks emphasise ICT-related competencies as separate domains (P21 and ATCS), others exhibit more integrative approaches where ICT skills are embedded within other competencies, such as critical thinking, problem-solving, communication, and collaboration—this is clearly the case for the NETS/ISTE framework (Joynes, 2019).

4.4 The Challenge of Addressing Transversal Competencies within the Curriculum

Studies indicate that many national curricula have moved towards integrating Transversal Competencies (European Commission 2002, OECD 2004, Hipkins, Boyd & Joyce 2005, Voogt & Pelgrum 2005). In their separate review of theoretical frameworks for Transversal Competencies, Choo and Villanueva (2012), Voogt

(2012) and UNESCO (2015) all identified a number of challenges for curriculum developers, particularly how to identify what competencies are actually new, how to define them, and how to embed them into existing curricula.

Education bodies that wish to embrace this new imperative have to decide whether to adopt an existing framework or to create their own. While it has perhaps been easier to identify what these competencies are by creating artificial frameworks, a greater challenge is to see where or if they fit within existing curricula and, if not, how this can be addressed by the educational bodies responsible for a country's curriculum development. The underpinning educational philosophies that form the foundation of a country's curriculum may or may not be closely aligned with the perceived importance of these new skills identifies in each framework.

UNESCO, in collaboration with ERI-Net, examined how ten different countries and economies define and apply non-academic skills in their education policies, practices, and curriculum frameworks (Brown, 2020). The findings suggest that different education systems utilise different methods of integrating the teaching and learning of Transversal Competencies into the curriculum. What they also provided was a framework which clustered a range of competencies in five categories; hence, suggesting that an arbitrary name for the framework is less important than the more granular recognition of specific competencies. The identification of specific skills provides an opportunity for a school system to focus on addressing these competencies within their own curriculum documents or educational programs rather than trying to align with a complete framework. Table 4.2 illustrates how the most commonly identified competencies were clustered into domains.

UNESCO also identified two main approaches to addressing Transversal Competencies: an *analytic approach*, where learning of TVCs is facilitated through a cluster

Table 4.2 © UNESCOs transversal competency framework

UNESCO tVCs	Key skills and competencies
Critical innovative thinking	creativity, entrepreneurship, resourcefulness, application skills, reflective thinking, decision making
Interpersonal skills	presentation skills, communication skills, leadership, organisational skills, teamwork, collaboration, initiative, sociability, collegiality, empathy, compassion
Intrapersonal skills	self-discipline, independent learning, flexibility, adaptability, self-awareness, perseverance, self-motivation, compassion, integrity, risk-taking, self-respect
Global citizenship	awareness, tolerance, openness, respect for diversity, intercultural understanding, conflict resolution, civic / political participation, respect for the environment, national identity
Media and information literacy	accessing information, locating information, communicating ideas, participating in democratic processes, analysing information and media, evaluating information and media content

Table 4.3 Ways of ensuring transversal competences are evident in the curriculum

Voogt (2012)	© UNESCO (2015)
TVCs be part of a new curriculum in which the traditional structure of school subjects is transformed, and schools are regarded as learning organisations (Plomp, 2013)	Specific subject: Learning of Transversal Competencies is included as a well-defined entity within the formal curriculum, for example, a subject with specific goals and syllabuses for formal teaching
TVCs be integrated as cross-curricular competences that both underpin school subjects and place emphasis on the acquisition of wider key competences	Extra-curricular: Learning of Transversal Competencies is made part of school life and embedded purposefully in all types of non-classroom activities
TVCs be added to the already existing curriculum as new subjects or as new content within traditional subjects	Cross-subject: Learning of Transversal Competencies runs across, infiltrates and/or underpins all 'vertical subjects', i.e. traditional school subjects

Table 4.4 Challenges to the implementation of transversal competencies into the curriculum

Definitional	Operational	Systemic
The competencies are poorly defined in the policy documents and often multiple terms are used and there is no indication of how the data can be assessed	Lack of teaching and assessment examples. Lack of training for teachers. No assessment or reflection opportunities provided to improve practice. Not a priority for budget allocations	With large class sizes and an overcrowded curriculum, there is little time to review and give opportunities for TVCs to be considered or given priority. There is little value given to the achievement of these skills

of learning areas or learning experiences, each intended to provide the learner with a particular competency; and a *holistic approach,* where Transversal Competencies are introduced as a scheme or program that conveys an overall message (McIlvenny, Sheffield, 2019). Excluding Japan, participating countries described in this study were found to use the analytic approach (© UNESCO, 2018). The two models that describe modes that Transversal Competencies could be included in the curriculum are outlined in Table 4.3.

The UNESCO Report (© UNESCO, 2015) identified several challenges facing educational institutions in their efforts to address Transversal Competencies and identified them as definitional, operational, and systemic. See Table 4.3 for a summary (Table 4.4).

TVC Assessment Challenges

The Australian and Indian curriculums are both currently undergoing review; however, there is still little emphasis on ensuring Transversal Competencies are strategically taught and assessed in both. Most core subjects such as science and mathematics focus on content knowledge and skills (both of which are relatively easy for teachers to assess) unlike the TVCs/General Capabilities which are not currently reported on in most curricula. The other issue is that these are complex competencies (problem-solving, critical thinking, and communication) which require students to complete multifaceted tasks applied to real-world situations with teachers needing to make judgments and discern evidence in complex interactive situations (Hipkins et al. 2005, Dede 2010). Learners must also be presented with the opportunity to practise and then to demonstrate these competencies across multiple settings and diverse situations (Hipkins et al. 2005). This is challenging in most educational settings as there is a rigid and inflexible timetable structure and in the secondary schools the learning is subject-siloed.

4.5 How Makerspaces Support the Development of Transversal Competencies

In Chapters One, Two, and Three, we have attempted to describe the characteristics, affordances, and benefits of having a Makerspace in the school. Makerspaces programs can provide a solution to many of the issues identified in this chapter around the teaching and development of Transversal Competencies in the classroom. Depending on the structure and focus of a school's curriculum, an individual or group of teachers might address some of the larger systemic issues identified in Table 4.7 as well as beginning to strategically focus on the development of some or all of these Transversal Competencies. It is through this imperative to address the development of Transversal Competencies that Makerspaces gain credibility and create opportunity. Makerspaces can provide the platform for and mechanism by which teachers explore how to teach Transversal Competencies, what they might look like and how evidence of their achievement can be monitored and assessed. It could potentially be the Action Research Hub in the school.

Our aim in Chapters Five, Six, Seven, and Eight is to explore in depth four Transversal Competencies and determine how a Makerspace may be used as a platform to develop and measure them. The chosen TVCs are higher-order thinking (Chap. 5), collaboration (Chap. 6), communication (Chap. 7), and resilience (Chap. 8).

4.6 Summary

What are Transversal Competencies?

Transversal Competencies describe a broad-based set of skills, knowledge, and understandings considered essential to live and work in the twenty-first century. They are also sometimes referred to as twenty-first-century, soft, or entrepreneurial skills.

Why are they important to develop in students?
Effective Transversal Competencies, integrated into lifelong learning processes and systems, are critical for both organisations and individuals. These competencies are considered increasingly more important in today's educational, organisational and economic forums. Students need to become more flexible, adaptable and agile in these times of uncertainty and change.

What do these skills look like?
There are many frameworks developed that attempt to describe these competencies. UNESCO has categorised them into five domains:

1. Critical innovative thinking;
2. Interpersonal skills;
3. Intrapersonal skills;
4. Global citizenship, and;
5. Media and information literacy.

Within each of these domains there are a number of competencies included. See Table 4.2 for details.

How can these skills be taught in a Makerspace?
To allow these competencies to be developed within a Makerspace, students need to be given the opportunity to think imaginatively, take risks, feel confident to try new things, collaborate with others to solve problems, learn from their mistakes and communicate through multiple mediums. A culture of 'making', designing, innovating, collaborating, and building, which is central to the underlying premise of Makerspace, provides the perfect context in which to undertake such activities. The acquisition of these skills is focused on more deeply in Sect. 2 in Chaps. 5–9.

Chapter 5
Higher-Order Thinking in a Makerspace

If we are not prepared to think for ourselves, and to make the effort to learn how to do this well, we will always be in danger of becoming slaves to the ideas and values of others due to our ignorance. (William Hughes)

Keywords Critical thinking · Creative thinking · Divergent thinking · Convergent thinking · Metacognition

Definition of Higher-Order Thinking

Higher-order thinking relates to the use of complex cognitive skills. These include critical thinking, creative thinking, and problem-solving that challenges one to 'think outside the box'.

Focus Questions

- What is higher-order thinking?
- Why is higher-order thinking important in a Makerspace?
- What is the difference between critical and creative thinking?
- How is higher-order thinking used to solve problems?

R. B. Koul et al., *Teaching 21st Century Skills*,
https://doi.org/10.1007/978-981-16-4361-3_5

5.1 Introduction

Higher-order thinking skills (HOTS) are a collection of cognitive processes used to make sense of, and create change in, the world around us. Within an educational setting, many of these processes have been compartmentalised and labelled to make it easier for educators to identify their characteristics and to develop them within a classroom context. Teachers, in their lesson delivery, are guided by how these skills have been represented within curriculum documents, as well as the popularised ways in which educational communities describe and promote them. For example, high profile taxonomies and frameworks such as Bloom's *Taxonomy of Learning*, Costa's *Habits of Mind* (2009), *The Four C's* (of which Critical and Creative Thinking are two of the 'C's') (Tsekeris, 2019), and Ryan's *Thinkers Keys* (2014) all have, at their core, the purpose of naming, organising, and operationalising the teaching of these skills. Along with the current imperative for higher-order thinking skills to be developed as an essential Transversal Competency (see Chap. 4 and Table 5.1) and as part of the *design thinking process* in STEM education, pressure is being placed on educators to demonstrate that they are now strategically teaching this nebulous set of skills in their classrooms.

In this chapter, we will provide a lens through which to examine higher-order thinking skills in relation to a Makerspace application. We have chosen to focus on three key concepts—**critical thinking, creative thinking** and **problem-solving**—which are defined as one of UNESCOs TVCs in Table 5.1. They were chosen due to their increasing presence in curriculum documents and the range of materials available to assist teachers in their development. Each of these concepts will be examined both separately and collectively, allowing teachers the opportunity to identify the distinctive elements of each and the dynamic, organic way they fit together. We will also identify strategies that will help you create a 'thinking classroom'.

Table 5.1 UNESCO's transversal competencies highlighting critical innovative thinking (adapted from Chap. 4, Table 4.2 - UNESCO's Transversal Competency Framework (UNESCO 2017))

UNESCO TVC	Key skills and competencies
Critical innovative thinking	**creativity,** entrepreneurship, resourcefulness, application skills, **reflective thinking, decision making, problem-solving, innovative thinking, critical thinking**
Interpersonal skills	presentation skills, communication skills, leadership, organisational skills, teamwork, collaboration, initiative, collegiality, sociability, empathy, compassion
Intrapersonal skills	self-discipline, self-respect, independent learning, flexibility, adaptability, self-awareness, resilience, perseverance, self-motivation, compassion, integrity, risk-taking, sense of belonging
Global citizenship	awareness, tolerance, openness, respect for diversity, intercultural understanding conflict resolution, civic / political participation, respect for the environment, national identity

5.2 Thinking and Maslow's Hierarchy of Cognitive Needs

Along with Maslow's original *Hierarchy of Basic Needs,* he developed a second hierarchy which dealt with the area of cognition. He believed that while our *need to know* drove us deep into the details of why things work, he also suggested that just knowing things was never enough: that we also need to *understand and make meaning* from what we discover. The ability and desire to think critically and creatively therefore lies at the heart of this need to know and understand (Fig. 5.1).

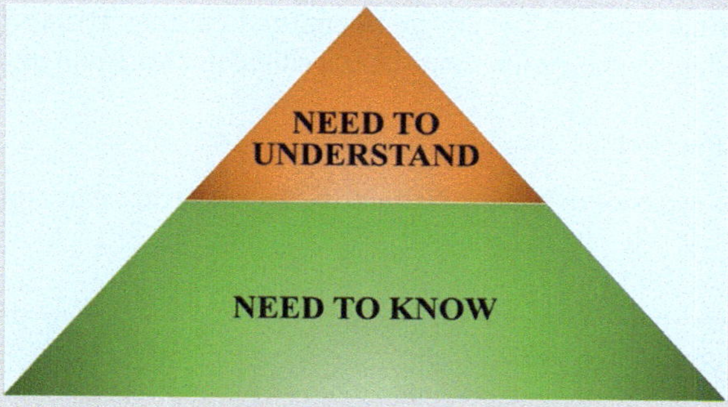

Fig. 5.1 Maslow's Hierarchy of Cognitive Needs

5.3 Critical and Creative Thinking

The P21 Partnership for Twenty-first-century Learning (2015) identifies critical and creative thinking as two of its learning and innovation skills which 'are increasingly being recognised as those that separate students who are prepared for a more complex life and work environment in the twenty-first century, and those who are not. A focus on creativity [and] critical thinking is essential to prepare students for the future'.

While often described together, critical and creative thinking are cognitive processes that could be seen as being at opposite ends of a **thinking spectrum**. Creative thinking refers to the formulation of original and resourceful ideas to innovate, while critical thinking is denoted by the use of one's intellect to discern, distinguish, evaluate, and reflect when decision making. Below we distinguish critical from creative thinking before describing the synergistic relationship between the two.

Critical Thinking

> "Critical thinking is at the core of most intellectual activity that involves students learning to recognise or develop an argument, use evidence in support of that argument, draw reasoned conclusions, and use the information to solve problems." (© Australian Curriculum and Reporting Authority [ACARA], 2020).

The word 'critical' comes from the Greek word 'kritikos', meaning discerning. Critical thinking delves deeper than mere observational thought to look beyond the superficial layers of something by questioning, analysing, and evaluating it. The term *convergent thinking* is often used to describe critical thinking (honing, looking under the microscope, deconstructing) as a detailed interrogation of something in order to make sense of it or possibly make it better.

The P21 Partnership for Twenty-first-century Learning Framework (2015) is widely used to identify the various aspects of critical thinking and describes them in four domains:

Reasoning Effectively

- Using various types of reasoning (inductive, deductive, etc.) as appropriate to the situation.

Using Systems Thinking

- Analysing how parts of a whole interact with each other to produce overall outcomes in complex systems.

Making Judgements and Decisions

- Analysing and evaluating evidence, arguments, claims, and beliefs;
- Analysing and evaluating major alternative points of view;
- Synthesising and making connections between information and arguments;
- Interpreting information and drawing conclusions based on the best analysis; and
- Reflecting critically on learning experiences and processes.

Solving Problems

- Solving different kinds of non-familiar problems in both conventional and innovative ways; and
- Identifying and asking significant questions that clarify various points of view and lead to better solutions (Ruutman, 2019).

Creative Thinking

You can't just give someone a creativity injection. You have to create an environment for curiosity and a way to encourage people and get the best out of them.

Sir Ken Robinson

Creative thinking is a mental process that generates a wide range of ideas and opens up opportunities. Often seen as the source of innovation and invention, creative thinking goes beyond what is *known* to what is *possible*.

The Australian Curriculum's *Critical and Creative Thinking General Capability* suggests that 'creative thinking involves students learning to generate and apply new ideas in specific contexts, seeing existing situations in a new way, identifying alternative explanations, and seeing or making new links that generate a positive outcome. This includes combining parts to form something original, sifting and refining ideas to discover possibilities, constructing theories and objects, and acting on intuition' (© Australian Curriculum and Reporting Authority [ACARA], 2020).

Creative thinking, often described as divergent thinking, results in original and constructive ideas. It can occur in two modes: unconscious (random) or conscious (reflective). Creativity often comes when we least expect it or are not even thinking about anything specific. This can occur while walking, eating, meditating, or at an unexpected moment, like Archimedes shouting 'eureka!' in a bathtub after discovering a method of determining the purity of gold. This is an example of unconscious creative thinking in which creative ideas occur unconsciously and spontaneously. A more conscious or reflective thinking process would be ruminating on a problem and conceiving one or multiple creative solutions. Creative thinking is a skill that can be developed and often relies on intuition, experience, and understanding to draw connections between seemingly dissimilar ideas or concepts. Design thinking and innovation are also often considered creative processes.

As with critical thinking, the P21 Framework also identifies three domains for developing creativity in the classroom:

Thinking Creatively

- Using a wide range of idea creation techniques (such as brainstorming, Thinker's Keys, etc.);
- Creating new and worthwhile ideas (both incremental and radical concepts); and
- Elaborating, refining, analysing, and evaluating original ideas to improve and maximise creative efforts.

Working Creatively with Others

- Developing, implementing, and communicating new ideas to others;
- Being open and responsive to new and diverse perspectives; incorporating group input and feedback into the work;
- Demonstrating originality and inventiveness in work and understanding the real-world limits to adopting new ideas;
- Viewing failure as an opportunity to learn; and
- Understanding that creativity and innovation are part of a long-term, cyclical process of small successes and frequent mistakes (The Partnership for Twenty-first-century Skills, 2020).

Implementing Innovation

- Acting on creative ideas to make a tangible and useful contribution to the field in which the innovation will occur.

Another environmental factor directly related to creativity is the reward system. Amabile (Kohn, 1987) has developed a Motivation Principle of Creativity and suggests that most students are creative when they feel motivated primarily by the interest, enjoyment, satisfaction and challenge of the work itself and not by external pressures. By creating challenging activities where students are given autonomy to decide the learning pathway, they are more likely to enjoy the process.

A 'Thinking' Synergy

Although we have separated critical and creative thinking, both processes are often simultaneously at work. These complementary skills can be used simultaneously to successfully solve problems. Scriven supports this idea by reinforcing the symbiotic relationship between the two—'critical skills go hand in hand with creative ones' (1979). Pauls and Elder also stated 'sound thinking requires both imagination and intellectual standards. When an individual engages in high-quality thinking, they are functioning both critically and creatively; producing and assessing; and generating and judging the products of his or her thought' (2006).

5.4 Using Blooms' Taxonomy as a Framework for Developing Higher-Order Thinking

Bloom's Taxonomy of Learning (Figure 5.2) is perhaps the most widely accepted and used framework in educational settings to scaffold the teaching of thinking skills. The taxonomy is represented as a hierarchy that distinguishes between lower- and higher-order thinking. Bloom (1956) suggests that 'lower-order' thinking skills include knowledge retention and application. Activities that utilise these skills include recounting facts, figures, and dates and arranging this information to determine outcomes and solutions. In contrast to this, higher-order thinking skills are defined as skills requiring analysis, synthesis, and evaluation. These cognitive skills include

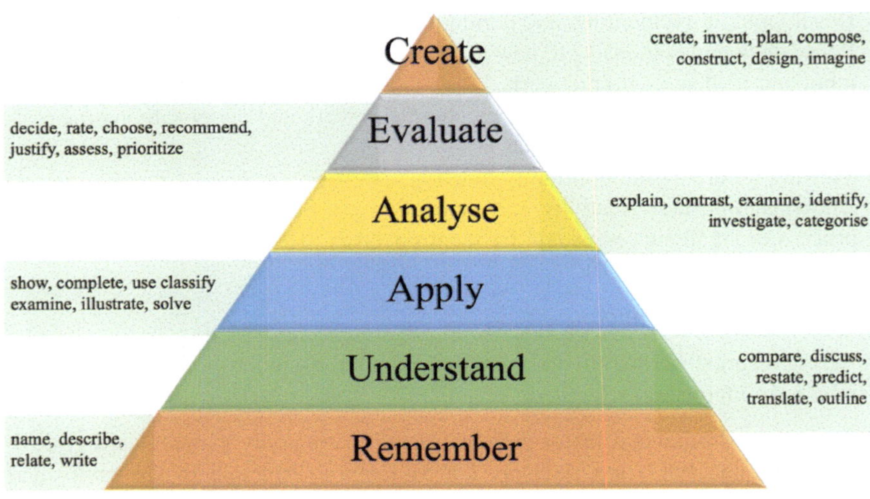

Fig. 5.2 Bloom's Taxonomy of Learning

categorising, classifying, and comparing and contrasting information in order to make decisions. Other characteristics include combining, creating, designing, developing, evaluating, and justifying decisions.

The terminology used and nature of this hierarchy (higher versus lower) may suggest that one set of skills is more important than the other, and that in order to reach the top, students must make their way through the lower levels first. We suggest that the hierarchy is an artificial construct, is not developmental in nature, and that skills are not taught sequentially but concurrently. They are determined by the nature of the task and context in which learners find themselves. In any learning context, students will be engaging in multiple thinking processes at once, whether conducting an inquiry, solving a problem, or innovating a new product. While the focus of this chapter is on **higher-order thinking** skills, we acknowledge the necessity and importance of developing thinking skills across all levels of the hierarchy. Educators need to be more organic in the way that they create opportunities for students to develop these skills. A Makerspace is an ideal environment to develop multiple thinking skills simultaneously through the learning process of making.

5.5 Creating a Thinking Classroom/Makerspace

Although interdependent, if we accept that critical and creative thinking are equally valid and necessary in the classroom, educators will be able to effectively employ teaching strategies that foster the development of both types of thinking. This does not, however, need to be done separately. Consider the idea of creating a 'thinking classroom' or 'thinking culture' which nurtures and promotes both deep interrogation and analysis, as well as creative thought, innovation, and problem-solving. Let the context and learning outcomes determine which thinking skills are required rather than artificially trying to teach a specific skill set.

To do this, we need to feel comfortable with letting go of control of the classroom and program to allow students to forge their own path. Sternberg (1987) implores educators to ask some hard questions: 'Do we really want students to think? Do we want them to become more critical and more questioning and less likely to accept things at face value?'. This challenges the very foundation on which educators develop philosophical viewpoints about what constitutes teaching and learning and whether we can encourage students to forge their own learning pathways.

Acknowledgement that there may be more than one 'right' answer to a problem, but instead, a complex array of appropriate yet diverse solutions to a problem and not merely one 'right' answer is another important step for educators to take when developing critical and creative thinking. Sternberg (1987) claims, 'very often in critical thinking problems there are no right answers. And even when there are, it is the thought process that counts'. A focus on process rather than outcome(s) will further aid in the application of higher-order thinking in the classroom. We can give students 'wicked' or ill-defined problems to solve or let them create the problems themselves.

Fig. 5.3 Thinking classroom

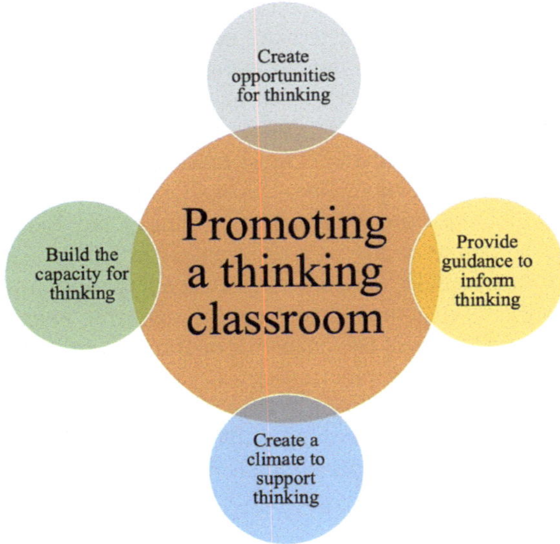

Another essential component in integrating higher-order thinking in the classroom is the capacity to instil self-efficacy in students. Sternberg (1987) realises that students 'must teach themselves, and all we can do is provide every possible means to enable this self-instruction to take place'.

Teachers that choose to embrace the maker philosophy, if not already present within their teaching philosophy and pedagogical repertoire, need to develop these attitudes. The following questions and Thinking Classroom Model (Figure 5.3 and Table 5.2) can help teachers identify the essential elements needed if they wish to create their own thinking classroom or Makerspace.

- How can I model higher-order thinking (HOT) for my students?
- What kind of learning environment is necessary to develop HOT in my classroom?
- What could I do to make HOT more intentional and purposeful in my classroom?
- How can I encourage students to be better thinkers and problem-solvers?
- How can I and my colleagues work collectively to prioritise effective higher-order thinking pedagogy across classrooms?
- How can I encourage creativity and innovation in my classroom and lesson plans?
- How can I encourage students to be more creative and innovative (Thomas, 2009)?

Table 5.2 Creating the thinking classroom

Classroom climate to support thinking	What does it look like?
Develop a non-threatening environment which encourages risk-taking, and 'thinking outside the box'.	Posters that promote risk-taking Solving wicked or ill-defined problems where the outcome is uncertain and there is no guarantee that a solution will be found Group work and peer assessment
Develop a questioning culture that uses different types of questions to frame engaging and thought-provoking tasks and reflections.	Questioning Frameworks and templates: 5Ws, Question Matrix; Six Thinking Hats; Bloom's Taxonomy of Questions; Socratic Questioning
Foster helpful *Habits of Mind* such as being attentive to detail during hands-on experiments.	Use Art Costa's 16 *Habits of Mind* and apply them strategically in the classroom when appropriate. Have a focus per week or month. Use posters and annotate worksheets with icons
Develop an environment where students are empowered and encouraged to take charge of their learning and make suggestions for how they solve a problem.	Hackerthons, STEM Fair, STEM Expo Group projects Makerspaces
Develop a problem-solving culture where students are required to use a range of critical thinking strategies to establish a solution.	Problem-solving process Problem-based learning Design process Inquiry process
Provide a resource rich learning environment (scaffolding and support).	Have multiple resource formats for students to access (books, posters, how-to videos, laminated instructions, kits, models)
Opportunities for thinking	*What does it look like?*
Provide multiple opportunities for creative processes such as brainstorming, mind mapping and ideation with no limits and no judgements about the ideas they produce. This will encourage students to take risks with their thinking (as there is no judgement and all ideas are accepted).	Mind-mapping templates and tools Writing walls Post-It notes Laminated copies of Thinker's Keys, Graphic Organisers Question Matrix Six Thinking Hats
Present problem-solving opportunities for which students have no predetermined solutions and for which there are more than one reasonable solution. (ill-defined or 'wicked' problems).	Problem-based learning Ill-defined problems Wicked problems
Encourage iterative thinking where students revisit initial ideas and share their evolving thinking through conversations, class discussions and written reflections. Can include designing a product to meet users' needs, pitching ideas to a prospective audience, seeking feedback to improve a solution.	*Rework the piece* Rework a product given a new audience, perspective or piece of information *Authentic Designs* Design a boardgame to support a Year 3 students' maths learning

(continued)

Table 5.2 (continued)

Classroom climate to support thinking	What does it look like?
Explicitly teach thinking strategies such as how to evaluate data or how to represent cause and effect relationships. Embed thinking processes into the curriculum.	Curriculum Documents (e.g. Critical and Creative Thinking General Capability—Australian Curriculum)
Create conditions that encourage students to actively engage in learning through critical inquiry.	Deep dives Inquiry learning Under the iceberg Curly questions Unanswerable questions Wicked problems Ill-defined problems
Ensure opportunities for student to design and invent using a range of strategies.	Design thinking process Robotics and physical computing Coding AR / VR
Emphasise reality by using authentic learning tasks which link 'real' audience needs in a school and local community and address local, national or global issues.	United Nations Sustainable Development Goals Targets—link to local context. Global Collaborative Projects
Create a culture of exploration and innovation in the classroom. A pioneering spirit will encourage innovation and trailblazing which are the foundation on which creativity is built.	Engage with real-world problem-solving where students can see their work make a difference to other people's lives.
Building Capacity to Think	*What does it look like?*
Provide opportunities for sustained inquiry through use of learning provocations and standard frameworks to scaffold the learning process.	Design thinking Guided inquiry Problem-based learning Hackerthon Trevor Mackenzie Inquiry provocations
Develop metacognitive processes by encouraging students to offer sound reasons or explain their thinking. (Why do you think that?).	Make thinking visible in the classroom by using posters, charts, graphic organisers etc. Incorporate 'thinking language' in the classroom Use Socratic Questions, Six Thinking Hats, SWOT Analysis, Habits of Mind
Co-construct learning and assessment criteria with students. Use these criteria to make judgements and justify choices made	Essential/desirable Better/best Not yet vs Mastery Option 1/Option 2 Hattie's Visible Learning Strategies Self-assessment against negotiated criteria
Creating the Conversation	*What does it look like?*

(continued)

Table 5.2 (continued)

Classroom climate to support thinking	What does it look like?
Develop key vocabulary related to thinking to help students recognise important distinctions and offer more precise observations and conclusions.	Word walls Thinking vocabulary lists Word clouds
Provide resources that promote manipulation and tactile experiences (hands-on activities)	Kits, laminated templates (e.g. Six Thinking Hats) Cardboard and construction materials, LEGO bricks (thinking with your hands)
Provide opportunities for students to develop and use a repertoire of thinking tools independently with growing confidence.	Thinkers Keys Six Thinking Hats Bloom's Taxonomy Question Frameworks Graphic Organisers
Facilitate activities that nurture creativity such as: designing, constructing, planning, producing, inventing, devising, making, building, programming, filming, animating, blogging, mixing, innovating, publishing, podcasting, directing / producing, writing, composing.	Provide resources, materials, templates, etc. that promote each of these activities
Providing Guidance to Inform Thinking	*What does it look like?*
Develop guidelines with students for working collaboratively in active, hands-on projects. ('How can we work together to find a solution to this problem?')	*Design to specs* Develop a product that meets given criteria Create online modules with downloadable instructions 3D printing
Reflective Practice Draw the class together after a problem-solving session to share, discuss and analyse the various approaches they used. ('What have we found out?')	Socratic dialogue SWOT Analysis Six Thinking Hats PMI
Help develop stronger competence with a range of *Process Models* that incorporate multiple thinking skills. For example, the *Inquiry Process* includes a series of skills including: initiating and planning, performing and recording, analysing and interpreting, and communicating. Each of these competencies is strengthened when students are supported in thinking critically in every aspect of their learning.	Inquiry process Problem-solving model Design thinking Project-based learning
Provide opportunities to make thinking visible through activities, artefacts and discussions.	Mind maps Learning journals Celebrations of learning Hackerthons Develop prototypes of design ideas

5.6 The Power of Questioning

One of the most powerful strategies that can be used in the classroom to develop higher-order thinking is that of questioning. The questions we ask as teachers and the frameworks we provide for students to generate their own questions and conceptual understandings are two powerful ways to promote a range of higher-order thinking skills. Table 5.3 provides some questions starters and behavioural verbs that can help educators to create learning opportunities that encourage higher-order thinking. They have been arranged to reflect Bloom's Taxonomy of Learning.

It is important that teachers also use and model the language of thinking—this is an essential part of developing metacognition, which is not only thinking about one's thinking but also being able to verbalise, explain, and describe it to others.

By using a common language, students can recognise the skill they are exercising and the level of complexity of a question. When they see words like 'define', 'recognise', 'recall', 'identify', 'label', 'understand', 'examine', or 'collect', they know they are being asked to recall facts and demonstrate their knowledge of content. When they see words like, 'apply', 'solve', 'experiment', 'show', or 'predict', they understand they are being asked to demonstrate application. When a task begins with 'appraise', 'judge', 'criticise', or 'decide', they understand the higher-order thinking skill they are practising is 'evaluation'. This is why taxonomies such as Bloom's are widely used as they provide a common language for the various cognitive processes educators are trying to teach.

5.7 Humour and Higher-Order Thinking

Being in a positive state of mind can have an impact on how students' produce creative work. Isen (Russo, 1986) found that student subjects put into a cheerful mood were more creative and more successful at problem-solving. She concluded that 'positive models influence creativity by changing the way cognitive material is organised… (B)eing happy may cue you into a large and richer cognitive content, and that could significantly affect your creativity'. The compatibility of humour and higher-order thinking is noted by Whitmer (1986): 'humour motivates, develops insight, expands thinking, and requires reactive critical reading. In the pleasurable, risk-free environments in which learning best takes place, humour provides opportunities to play with ideas'. One of Art Costa's Habits of Mind is that of 'finding humour' (2009). Teachers have the opportunity to integrate humour into their classroom to boost overall enjoyment and student comfortability by highlighting humorous stories, images, anecdotes, and educational materials. An additional step is using parts of appropriate humour as part of an educators' instructive or conversational discourse.

Table 5.3 Questions to support and elicit the development of higher-order thinking

Bloom's hierarchy level	Keywords	Questions
Level 1: Knowledge—exhibits previously learned material by recalling facts, terms, basic concepts and answers. *Interpreting, exemplifying, summarising, inferring, paraphrasing, classifying, comparing, explaining,*	who, what, why, when, omit, where, which, choose, find, how, define, label, show, spell, list, match, name, relate, tell, recall, select	What is? How is? Where is? When happened? How did ______ happen? How would you explain? Why did? How would you describe? When did? Can you recall? How would you show? Can you select? Who were the main? Can you list three? Which one? Who was?
Level 2: Comprehension—Demonstrating understanding of facts and ideas by organising, comparing, translating, interpreting, giving descriptions and stating main ideas.	compare, contrast, demonstrate, interpret, explain, extend, illustrate, infer, outline, relate, rephrase, translate, summarise, show, classify	How would you classify the type of? How would you compare? contrast? What is the main idea of? Which statements support? Can you explain what is happening? What is meant by? What can you say about? Which is the best answer? How would you summarise?
Level 3: Application—solving problems by applying acquired knowledge, facts, techniques and rules in a different way. *Implementing, applying, using, editing, combining,*	apply, build, choose, construct, develop, interview, make use of, organise, experiment with, plan, select, solve, utilise, model, identify	How would you use? What examples can you find to? How would you solve? How would you show your understanding of? What approach would you use to? How would you apply what you learned to develop? What other way would you plan to?

(continued)

Table 5.3 (continued)

Bloom's hierarchy level	Keywords	Questions
Level 4: Analysis—examining and breaking information into parts by identifying motives or causes; making inferences and finding evidence to support generalisations. *Comparing, organising, deconstructing, outlining, structuring, integrating,*	analyse, categorise, classify, compare, contrast, discover, dissect, divide, examine, inspect, simplify, survey, take part in, test for, distinguish, distinction, theme, relationships, function, motive, inference, assumption, conclusion	What are the parts or features of? How is _______ related to? Why do you think? What is the theme? What motive is there? What inference can you make? What conclusions can you draw? How would you classify? How would you categorise? What evidence can you find? What is the relationship between? Can you make a distinction between? What is the function of? What ideas justify?
Level 5: Synthesis—compiling information together in a different way by combining elements in a new pattern or proposing alternative solutions.	build, combine, compile, compose, construct, create, develop, estimate, formulate, invent, predict, propose, solve, solution, suppose, modify, change, original, improve, adapt, delete, theorise, elaborate, improve,	What changes would you make to solve? How would you improve? Can you propose an alternative? How would you adapt ________ to create a different? How could you change (modify) the plan? What could be done to minimise (maximise)? What could be combined to improve/change? How would you test? Can you formulate a theory for? Can you predict the outcome if?

(continued)

Table 5.3 (continued)

Bloom's hierarchy level	Keywords	Questions
Level 6: Evaluation—presenting and defending opinions by making judgements about information, validity of ideas or quality of work based on a set of criteria. *Checking, hypothesising, critiquing, experimenting, judging, testing, detecting, monitoring, reviewing,*	choose, conclude, criticise, decide, defend, determine, dispute, evaluate, judge, justify, measure, compare, rate, recommend, select, agree, interpret, explain, appraise, prioritise, opinion, support, importance, criteria, assess, perceive, value, estimate, deduct	What is your opinion of? Would it be better if? Why did they (the character) choose? What would you recommend? How would you rate the? How would you evaluate? How could you determine? How would you prioritise? What judgement would you make about? Based on what you know, how would you explain? What information would you use to support the view? How would you justify? What data was used to make the conclusion? Why was it better that?

5.8 The Problem-Solving Process

Problem-solving is defined in this book as a four-part process. It incorporates both critical and creative thinking and requires the execution of a series of steps:

- defining the problem;
- generating alternative solutions;
- evaluating and selecting the 'best' solution; and
- implementing the solution.

Figure 5.4 illustrates where critical and creative thinking are emphasised in the problem-solving process. They are both present in each of these stages (Define the problem, etc.) although there may be more emphasis in one stage depending on the context of the problem-solving activity. Whereas creative thinking is useful to expand our range of options through ideation, critical thinking allows us to clarify the problem and to analyse and test options. In the case of innovation, it must be viewed in terms of creating products which are useful, which requires evaluation.

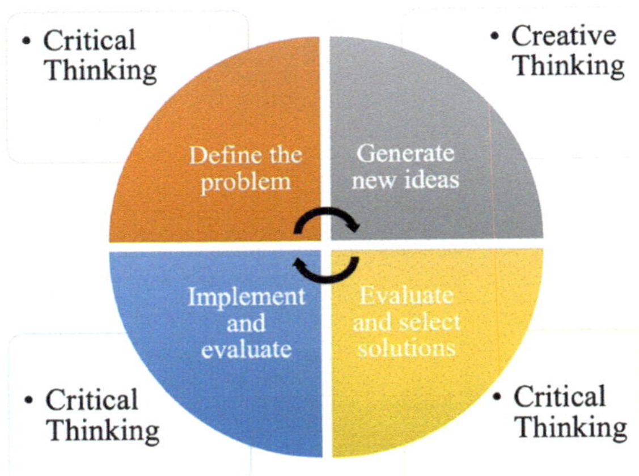

Fig. 5.4 Problem-solving process

5.9 Using Makerspace Design Process

In a Makerspace, students have access to resources and materials that can stimulate their thinking and creativity to solve a problem or develop innovative solutions. Graves (2014) states that 'we want our students to develop problem-solving skills and become engaged with making technology, not just using it'. Makerspaces address this need further by giving students direct experience in creative and design thinking, being able to use problem-solving skills while better preparing them to enter the twenty-first-century workforce. The design process used in Makerspaces guides makers.Many participants use a design thinking mindset to guide their making. Design thinking or the design process is a five-step model that includes: Ask, imagine, plan, create, and improve ('The Engineering Design Process', 2021). According to Mansback (2015), when planning, designing, constructing, and production skills are demonstrated by students, they are using a high level of critical thinking. Martinez and Stager (2013) also advocate Makerspaces as crucial to a student's engagement with a project as '[tinkering and making] are powerful ways to learn'. They also argue that students that are able to experience and engage in Makerspace activities first-hand are able to employ and demonstrate a variety of twenty-first-century skills.

5.10 Assessing Problem-Solving

A widely used framework for assessing individual students' problem-solving competence is the PISA Problem-Solving Framework (Zervas & Sampson, 2016). This framework provides a structure that allow teachers to identify some of the important thinking processes that can be implemented in a thinking classroom or a Makerspace. Collaborative problem-solving (CPS) is a critical and highly necessary skill used in education and the workforce. While problem-solving, as defined in PISA 2012 (OECD, 2010), relates to individuals working alone on solving problems where a method of solution is not immediately obvious, in CPS, individuals pool their understanding and effort and work together to solve these problems. The competencies assessed in the PISA 2015 collaborative problem-solving assessment therefore need to reflect the skills found in project-based school learning and in workplace collaboration as described above. In such settings, students are expected to be proficient in skill such as communicating, managing conflict, organising a team, building consensus, and managing progress.

Within the Australian Curriculum, critical and creative thinking are identified as a general capability. These skills have been scaffolded into a continuum that describes what skills are to be developed at each year level (© Australian Curriculum and Reporting Authority [ACARA], 2020). Table 5.4 illustrates how critical and creative thinking skills have been deconstructed into observable, assessable behaviours.

Case Study

The following case study provides a context to explore what creativity might look like as part of a Makerspace activity. Consider which observable behaviours from Table 5.4 the student is demonstrating.

Table 5.4 Critical and Creative Thinking General Capabilities Australian Curriculum (© Australian Curriculum and Reporting Authority [ACARA], 2020)

Sub-element	Level 1	Level 2	Level 3	Level 4	Level 5	Level 6
Inquiring – identifying, exploring and organising information and ideas element						
Pose questions	pose factual and exploratory questions based on personal interests and experiences	pose questions to identify and clarify issues, and compare information in their world	pose questions to expand their knowledge about the world	pose questions to clarify and interpret information and probe for causes and consequences	pose questions to *probe assumptions* and investigate complex issues	pose questions to *critically analyse* complex issues and abstract ideas
Identify and clarify information and ideas	identify and describe familiar information and ideas during a discussion or investigation	identify and explore information and ideas from source materials	identify main ideas and select and clarify information from a range of sources	identify and clarify relevant information and prioritise ideas	clarify information and ideas from texts or images when exploring challenging issues	clarify complex information and ideas drawn from a range of sources
Organise and process information	gather similar information or depictions from given sources	organise information based on similar or relevant ideas from several sources	collect, compare, and categorise facts and opinions found in a widening range of sources	*analyse*, condense and combine relevant information from multiple sources	*critically analyse* information and evidence according to criteria such as validity and relevance	*critically analyse* independently sourced information to determine bias and reliability
Generating ideas, possibilities and actions element						
Imagine possibilities and connect ideas	use imagination to view or create things in new ways and connect two things that seem different	build on what they know to create ideas and possibilities in ways that are new to them	expand on known ideas to create new and imaginative combinations	combine ideas in a variety of ways and from a range of sources to create new possibilities	draw parallels between known and new ideas to create new ways of achieving goals	create and connect complex ideas using imagery, analogies and symbolism

(continued)

Table 5.4 (continued)

Sub-element	Level 1	Level 2	Level 3	Level 4	Level 5	Level 6
Consider alternatives	suggest alternative and creative ways to approach a given situation or task	identify and compare creative ideas to think broadly about a given situation or problem	explore situations using creative thinking strategies to propose a range of alternatives	identify situations where current approaches do not work, *challenge existing ideas and generate alternative solutions*	generate alternatives and innovative solutions, and adapt ideas, including when information is limited or conflicting	speculate on creative options to modify ideas when circumstances change
Seek solutions and put ideas into action	predict what might happen in a given situation and when putting ideas into action	investigate options and predict possible outcomes when putting ideas into action	experiment with a range of options when seeking solutions and putting ideas into action	assess and test options to identify the most effective solution and to put ideas into action	predict possibilities, and identify and test consequences when seeking solutions and putting ideas into action	assess risks and explain contingencies, taking account of a range of perspectives, when seeking solutions and putting complex ideas into action
Reflecting on thinking and processes element						
Think about thinking (metacognition)	describe what they are thinking and give reasons why	describe the thinking strategies used in given situations and tasks	reflect on, explain and check the processes used to come to conclusions	reflect on assumptions made, consider reasonable criticism and adjust their thinking if necessary	assess assumptions in their thinking and invite alternative opinions	give reasons to support their thinking, and address opposing viewpoints and possible weaknesses in their own positions
Reflect on processes	identify the main elements of the steps in a thinking process	outline the details and sequence in a whole task and separate it into workable parts	identify pertinent information in an investigation and separate into smaller parts or ideas	identify and justify the thinking behind choices they have made	evaluate and justify the reasons behind choosing a particular problems-solving strategy	balance rational and irrational components of a complex or ambiguous problem to evaluate evidence

(continued)

Table 5.4 (continued)

Sub-element	Level 1	Level 2	Level 3	Level 4	Level 5	Level 6
Transfer knowledge into new contexts	connect information from one setting to another	use information from a previous experience to inform a new idea	transfer and apply information in one setting to enrich another	apply knowledge gained from one context to another unrelated context and identify new meaning	justify reasons for decisions when transferring information to similar and different contexts	identify, plan and justify transference of knowledge to new contexts
Analysing, synthesising and evaluating reasoning and procedures element						
Apply logic and reasoning	*identify the thinking used to solve problems* in given situations	*identify reasoning* used in choices or actions in specific situations	*identify and apply appropriate reasoning* and thinking strategies for particular outcomes	*assess* whether there is adequate reasoning and evidence to justify a claim, conclusion or outcome	*identify gaps in reasoning* and missing elements in information	*analyse* reasoning used in finding and applying solutions, and in choice of resources
Draw conclusions and design a course of action	share their thinking about possible	*identify alternative courses of action* or possible conclusions when presented with new information	draw on prior knowledge and use evidence when choosing a course of action or *drawing a conclusion*	*scrutinise ideas* or concepts, *test conclusions* and *modify actions* when designing a course of action	*differentiate the components* of a designed course of action and tolerate ambiguities when drawing conclusions	*use logical and abstract thinking* to *analyse and synthesise* complex information to info

(continued)

Table 5.4 (continued)

Sub-element	Level 1	Level 2	Level 3	Level 4	Level 5	Level 6
Evaluate procedures and outcomes	check whether they are satisfied with the outcome of tasks or actions	*evaluate* whether they have accomplished what they set out to achieve	*explain and justify* ideas and outcomes	*evaluate* the effectiveness of ideas, products, performances, methods and courses of action against given criteria	explain intentions and *justify ideas*, methods and courses of action, and account for expected and unexpected outcomes against criteria they have identified	*evaluate* the effectiveness of ideas, products and performances and implement courses of action to achieve desired outcomes against criteria they have identified

Demonstrating Creativity

Students that are high achieving in more directed class activities can still find the following open-ended activities challenging. In 2020, we created a Makerspace in a Primary School with a focus on a range of activities including terrariums, Bee hotels, Wigglebots, and pipelines activities. A high-achieving student in Year 6 found Makerspace activities challenging and throughout the maker weeks, he would ask 'Well, why don't you just tell me the right answer?'. He struggled with the uncertainty of the outcomes that are produced through the creative process.

Instructions

Use your 1L plastic bottle to

create a new item that your group agree to design and make (you might want to write your ideas down first) (Fig. 5.5).

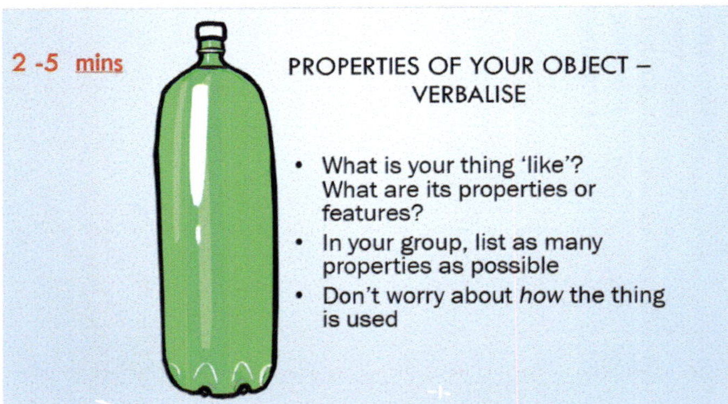

Fig. 5.5 Activity outline

Approaching the problem differently

Groups approached this task quite differently.

One group listed their combined diverse ideas: an elephant with a small trunk, a spaceship, and a garden bed. Another group of students wrote their ideas down on the page and then negotiated through each one until they reached consensus. A third group (not shown here) only wrote down one idea generated by one student and the other student just agreed with that idea, and they set about to make that object (Figs. 5.6 and 5.7).

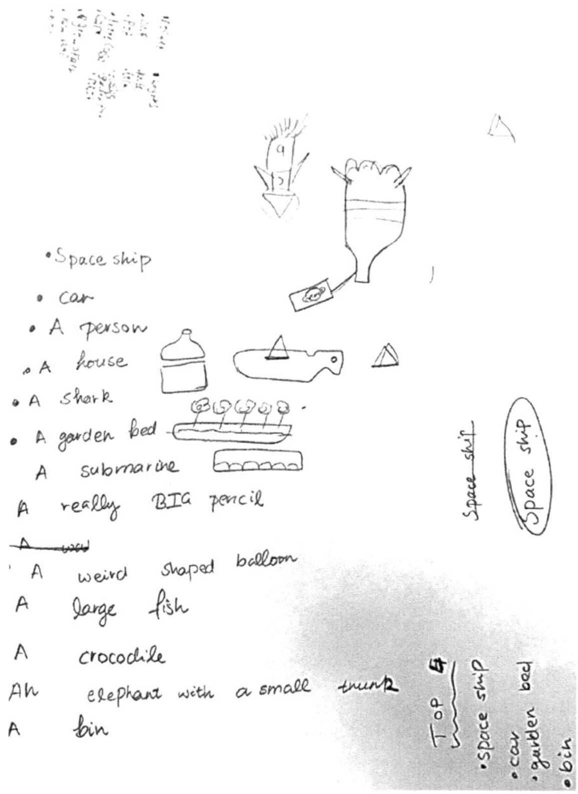

Fig. 5.6 Students listing ideas

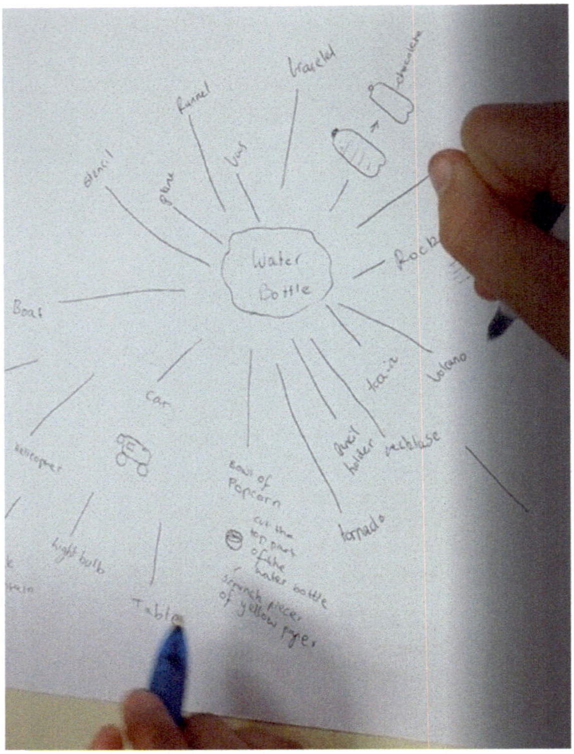

Fig. 5.7 Students listing ideas

The group pictured in Figure 5.8 had an unusual idea: they wanted to build the Leaning Tower of Pisa from their bottle and had considerable issues trying to get the bottle to lean. In the diagram, it can be seen that there is a lean, but it does not detail how they would ensure it was stable and leaning. This has been a concept that has challenged researchers in Pisa for centuries and not one that the Year 6s solved.

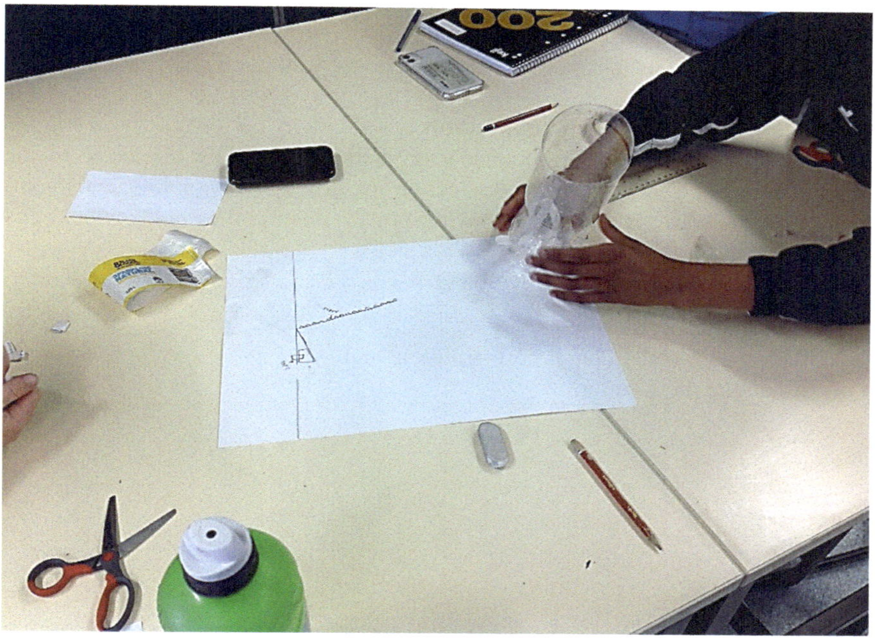

Fig. 5.8 Leaning Tower of Pisa example

Each group approached the task in significantly different ways based around their own understandings and experiences. Their alternative ideas and products highlighted the nature of creative thought and the critical thinking that was undertaken throughout the process.

Summary

Makerspaces can have a positive effect on how students use creativity and critical thinking skills to solve problems that are authentic and often ill-defined (McKay et al 2016). Learners who engage in hands-on and experimental learning experiences, like those of a Makerspace, tend to demonstrate the ability to take risks and produce innovative solutions.

What is higher-order thinking?

Higher-order thinking is a collection of cognitive processes that are used to navigate our way through life. It includes critical and creative thinking, problem-solving, and design thinking.

Why is higher-order thinking important in a Makerspace?

Makerspaces are designed to allow students to create, innovate and solve problems, therefore, they require a range of higher-order thinking skills to achieve this.

What is the difference between critical and creative thinking?
Critical and creative thinking are cognitive processes that could be seen as being at opposite ends of a *thinking spectrum*. Creative thinking refers to generating or using unusual ideas to innovate while critical thinking denotes evaluating, reflecting and judging information, insights, or ideas to make a decision.

How is higher-order thinking used to solve problems?
Problem-solving is the act of defining a problem; determining the cause of the problem; identifying, prioritising, and selecting alternatives for a solution; and implementing the solution. The problem-solving process incorporates both critical and creative thinking and includes the following steps:

- Defining the problem;
- Generating alternative solutions;
- Evaluating and selecting the 'best' solution; and
- Implementing the solution.

Appendix

Appendix 5.1

Habits of Mind

https://www.habitsofmindinstitute.org/

Persisting	Questioning and posing problems	Thinking and communicating with clarity and precision	Creating, imagining and innovative
Thinking flexibly	Responding with wonderment and awe	Remaining open to continuous learning	Thinking interdependently
Finding humour	Managing Impulsivity	Listening with understanding and empathy	Thinking about your thinking
Striving with accuracy	Applying past knowledge to new situations	Gathering data through all of your senses	Taking responsible risks

Appendix 5.2

Six Thinking Hats

https://www.habitsofmindinstitute.org/

WHITE HAT The White Hat symbolises facts and information	**YELLOW HAT** The Yellow Hat symbolises brightness and optimism. This hat focuses on positives and benefits	**BLACK HAT** The Yellow Hat symbolises judgement and focuses on negatives or what might go wrong
RED HAT The Red Hat symbolises emotions and feelings and looks at fears, likes, dislikes, loves and hates.	**GREEN HAT** The Green Hat symbolises creativity—looking at possibilities, alternatives and new ideas. It encourages new concepts and new perceptions.	**BLUE HAT** The Blue Hat symbolises the thinking process—metacognition or 'thinking about ones thinking'.

Appendix 5.3

Swot Analysis

STRENGTHS	WEAKNESSES	OPPORTUNITIES	THREATS
What are the strengths of this? What are the advantages of this? How can strengths be turned into opportunities?	What are the weakness of this? How could this be detrimental? What could stop this from being successful?	What opportunities can come from this? How will this make it better?	What will stop this from being successful?

Appendix 5.4

Thinkers Keys

https://www.thinkerskeys.com/

DECISIONS This strategy helps you to make a decision. Highly suitable for procrastinators.	**INFORMATION** The focus is on information to help answer questions or to undertake research.	**RUBRICS** Provides criteria that set the standards for a learning task.	**ACTION** Write out the Action Plan for doing a task.

(continued)

(continued)

DECISIONS This strategy helps you to make a decision. Highly suitable for procrastinators.	**INFORMATION** The focus is on information to help answer questions or to undertake research.	**RUBRICS** Provides criteria that set the standards for a learning task.	**ACTION** Write out the Action Plan for doing a task.
PERSPECTIVES Allows for different perspectives or viewpoints from which to explore a topic or issue.	**PURPOSE** Clarify your reasons for engaging in an activity.	**REFLECTION** A metacognitive process that allows you to think about your own thinking.	**CONSEQUENCES** What can or will occur as a result of an action taken or decision made.
QUESTION Use questions to direct the way you think about a topic. Start with the 5Ws.	**THREE WHYS** Ask 'Why is that?' three times over when you are exploring an issue.	**BRAINSTORMING** Generate many ideas about a topic	**COMBINATIONS** Combine disparate ideas or objects into an all-new ones.
IMPROVEMENT Design improvements to a specific product or process.	**CHALLENGE** Challenge yourself to think very differently by addressing a provocative statement.	**INVENTIONS** Develop innovative devices that could be used in your everyday life.	**BAR** Use the 3 steps of BAR (Bigger Add Replace) to make some changes to an object or idea
BRICK WALL Clarify what is stopping you from accomplishing something; and then think of how to get around it.	**IN COMMON** Name two different things / concepts, and then brainstorm common points between them.	**REVERSE** Generate answers to a reverse question, e.g. How can we encourage poor nutrition? And then list the opposite answers.	**PREDICTION** Think ahead. Ponder how things may have changed in 5, 10 or even 100 years from now.

Chapter 6
Developing Collaboration Skills in a Makerspace

Alone we can do so little; together we can do so much. (Helen Keller)

Keywords Collaborate · Share · Negotiate · Partner · Consensus · Shared understanding

Definition of Collaboration
Collaboration—the situation of two or more people working together to create or achieve the same thing (Cambridge Dictionary).

In schools, **collaborative learning** is the use of small groups where students work together to maximise their own and each other's learning.

Focus Questions

- What is collaboration?
- How can Makerspaces support collaboration?
- What is the BAR model and how can it help teachers create a collaborative space?
- What can teachers do to promote collaboration?

R. B. Koul et al., *Teaching 21st Century Skills*,
https://doi.org/10.1007/978-981-16-4361-3_6

6.1 Introduction

The concept of the collaborative classroom was popularised in the 80 s and 90 s and evolved from Vygotsky's Social Development Theory (1978), which highlighted the importance of learning through interactions with others rather than solely independent work. Vygotsky stressed that the fundamental role of social interaction in the development of cognition as he strongly believed that community plays a central role in the process of 'making meaning'.

The power of collaboration allows an individual's ideas to be presented to others and then reframed using different lenses to achieve a common goal through discussion and argument. This process creates a deeper, shared understanding of a problem or task to achieve an outcome or find a solution. In this format, collaboration creates a safe space for students to simultaneously challenge and develop in their thinking.

> "Collaboration is essential in our classrooms because it is inherent in the nature of how work is accomplished in our civic and workforce lives. Fifty years ago, much work was accomplished by individuals working alone, but not today. Much of all significant work is accomplished in teams, and in many cases, global teams."
>
> (NEA, pg. 20).

Vygotsky noted that children interacting towards a common goal tend to regulate each other's actions. Forman & Cazden (1986) also observed that when students work together on complex tasks, they assist each other in much the same way adults assist children—in such tasks, dialogue consists of **mutual regulation**. Together, they can solve difficult problems they could not otherwise solve when working independently (Tinzmann et al. 1990).

Collaboration allows multiple perspectives of a problem to be explored as everyone approaches a situation differently. It also provides the opportunity for a student's cognitive and emotional disposition to be challenged as they experience a range of emotions, perhaps not experienced while working independently. This is because they have to accommodate and empathise with others and examine ideas from another persons' perspective or viewpoint. These are all very important elements in developing a students' sense of self and also developing their resilience, especially if their ideas are challenged by others.

Collaboration and Maslow's Hierarchy of Needs

For collaboration to work effectively in a classroom setting, every member of the group needs to feel as if they have a role to play. Their contributions must feel valuable and will also help achieve a common goal or solve a specific problem. In terms of Maslow's hierarchy levels, three (Belonging) and four (Respect) are the needs being addressed (Fig. 6.1).

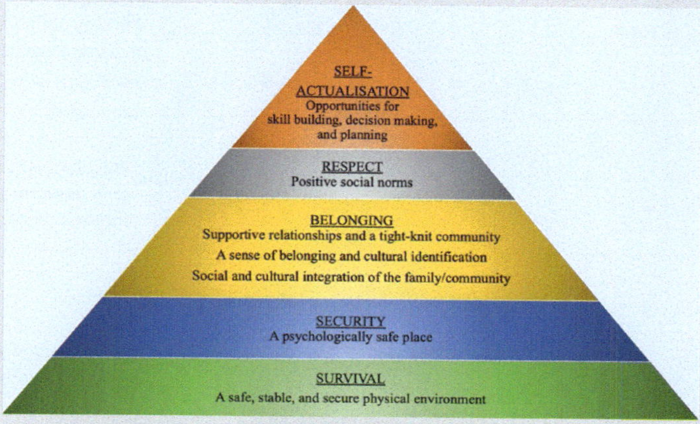

Fig. 6.1 Collaboration and Maslow's Hierarchy of Needs

There is no such thing as a self-made man. You will reach your goals only with the help of others.

George Shinn.

6.2 Collaboration as a Dynamic Process

If we accept that collaboration occurs to suit a specific need and context, then it could be considered as an event that has finite parameters—purpose, group members, time frame and desired outcomes. Collaboration does not, however, need to be strategically planned and executed. It can be both formal and informal in nature and can arise from a planned project or serendipitous meeting. The collaboration may be a five-minute discussion to clarify an idea between two students, a weeklong project to create a product, or a large collaboration of people from different locations around the world to solve a problem over several months. Within the collaborative event, however, there are dynamic processes continually occurring and organically evolving. It is useful to remember that collaboration can be synchronous (face to face/live), or asynchronous (online/global projects). We have developed the BAR model (Fig. 6.2) to help explain what the dynamic process of collaboration might look like within Makerspace context. Figure 6.2 describes a three-stage process that a collaborative experience would transition through:

- **Build**—Establishing the purpose and parameters of the collaboration (Why is this being done? How are we going to do it? What do we hope to achieve?)

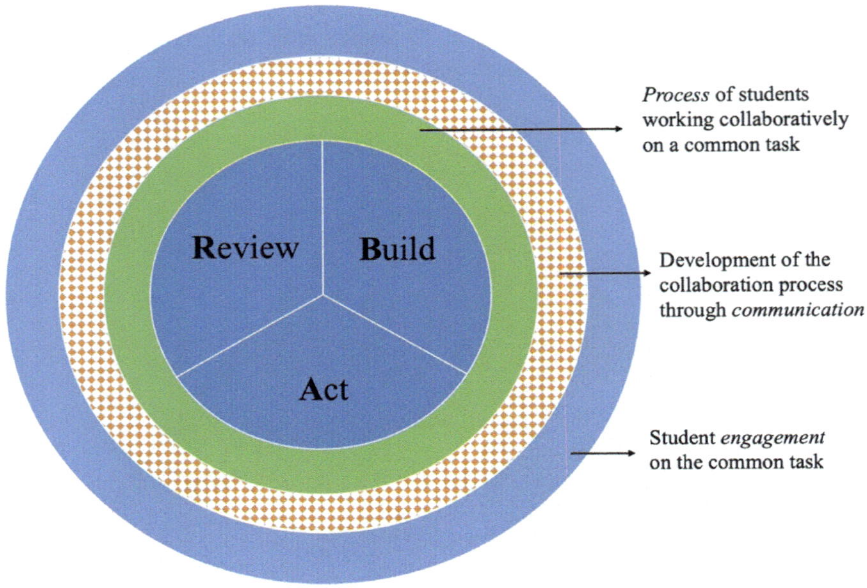

Fig. 6.2 Collaboration BAR model

- **Act**—Taking action to achieve the desired outcome (solving a problem or creating a product or performance)
- **Review**—Reflecting and evaluating the process undertaken and whether the desired outcomes were achieved.

At each of these three stages (or phases) of collaboration, we have identified two essential components we believe are necessary for collaboration to be considered effective. While they are not the only ones that are important, we have chosen to explore two:

- *Process*—the series of actions or steps taken in order to complete the collaboration successfully, and;
- *Efficacy*—the extent and effectiveness to which the individuals in the group engage in the collaborative process.

Table 6.1 deconstructs each of the collaborative stages into identifiable and observable actions that to monitor the effectiveness of the collaboration in your Makerspace.

The Collaborative Process versus a Child's Ability to Collaborate

- While the emphasis in this chapter is the collaborative process and how educators can create opportunities for students to participate in this process,

Table 6.1 Three domains of the collaborative process

BUILD	ACT	REVIEW
Process Identify the purpose, timeframe and desired outcomes of the collaboration Determine how the collaboration will be documented/recorded Identify steps to be undertaken throughout the collaboration Negotiate roles and responsibilities to group members Establish communication processes and formats to be used	**Process** Monitor group member actions to ensure roles are being adhered to Monitor groups' ability to follow agreed collaborative process Use agreed communication formats to record group actions	**Process** Evaluate the effectiveness of: the outcome of the collaboration the collaborative process the documentation process the communication strategies and formats used to document the process and disseminate the findings/outcomes of the collaboration
Efficacy Establish group norms/team rules Determine the purpose of the group/collaboration Determine 'Rules of Engagement' Individuals set personal goals for their engagement in the collaboration	**Efficacy** Monitor individual participation in allocated roles/activities Monitor learners attempts to use alternative strategies when required Monitor learners' ability to resolve differences according to the 'Rules of Engagement'	**Efficacy** Evaluate the effectiveness of: The established 'Rules of Engagement' The allocation of roles and responsibilities of each group member

there is an implicit understanding that students will develop a range of identifiable and assessable collaboration skills through a range of contexts over time.

Find a group of people who challenge and inspire you,
spend a lot of time with them, and it will change your life.
Amy Poehler.

6.3 Developing a Collaborative Makerspace Culture

Gutsche (2013) suggested that 'Makerspaces are not just about the physical act of fabricating things. They are **collaborative**, exploratory, community spaces that foster a spectrum of skills, which enable people to be fully engaged twenty-first-century citizens'. Burke (2014) contends that the maker movement is changing from an individual only mindset to one of collaboration of makers in the learning space. Martinez and Stager (2013) also cite collaboration as one of the most important twenty-first-century skills that can be developed within a maker environment. They note that 'when the collaboration is authentic, students will gain a greater appreciation for the benefits of collaborating and the result of the experience will be richer'. Most of the work and projects that students do in Makerspace are naturally collaborative or, at the very least, can be restructured to include different roles and responsibilities that lead to the chance for students to collaborate. Collaborative activities could include students working together on a project, mentoring others, and sharing information and materials (Clapp et al. 2017).

> None of us, ever do great things. But we can all do small things, with great love, and together we can do something wonderful.
>
> Mother Teresa.

Essential Elements of a Collaborative Makerspace
There are a number of essential elements that need to be included in order to create a robust collaborative Makerspace.

Groupings
By creating collaborative groups, teachers are able to strategically cluster students with varying ability levels and skill sets. There are many ways that student groups can be established, and this will very much depend on the nature of the class and the desired outcomes of the activities/program. Teachers may also allocate students to groups or instruct them to select their own, with either approach having benefits and limitations. Ideally, multiple group formats would be used over the course of the year. It may start with pair-groupings to establish structure and routines and then eventually scaling up to larger groups.

Group Norms/Team Rules
It is important that agreed procedures are developed under which the group will behave. These norms ideally should be identified by the group, so they have ownership of the accepted behaviours. It is important that educators guide their attempts at creating a set of group norms. It will help for the negotiated rules to be written and displayed in plain sight, or to have students keep their own copy of the rules in their working file (paper or digital) to remind them when working through a collaborative activity, for an example, refer back to Chap. 3, the Elmtree Primary School Bee Hotels Project, where written group norms were created to govern behaviour in a virtual space.

Learner Autonomy/Agency
The negotiated classroom provides an ideal setting for the development of autonomous learners where students are empowered to take responsibility for their own learning. This learning culture does, however, take time, careful consideration, and strategic development by the teacher. The gradual removal of constraints and the encouragement of students to 'step up and speak out' is most effectively done by establishing autonomy milestones that allow teachers to gauge when students

Fig. 6.3 Question funnel

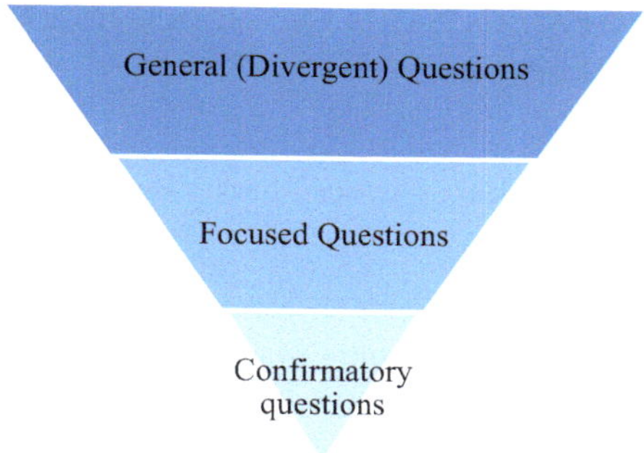

Essentials Skills Developed in a Collaborative Makerspace

are ready for the next level of freedom within the space. Students are taught how to be a good negotiator by listening well, showing patience and flexibility, pointing out shared ideas and areas of group agreement, and thinking under pressure. It also teaches them to put their point to their team respectfully when they disagree with others in the group.

Teacher as Mediator
As students gain more autonomy in their learning and collaborative groups begin to self-regulate their activities, the teacher becomes a 'guide on the side', intervening only when necessary to refocus a group or help them gain consensus for a decision.

Communication
Communication will be explored in greater detail in Chap. 7; however, two key communication elements are highlighted below to reinforce their essential place within collaboration.

Active Listening
Students should be encouraged to use active listening strategies including making eye contact, offering empathy, and refraining from cutting others off.

Effective Questioning This is central to many Makerspace activities and provides the opportunity for teachers to encourage students to question their actions and thinking. Questions can be used at any stage during the course of a lesson. Teachers may set the scene by asking focus questions which will guide the learning for the class activity. As students' progress through the activities, teachers may challenge their actions in a positive way by asking them to justify why they behaved or thought in a certain way. Student-initiated questioning also effectively deconstructs learning goals and outcomes and allows the examination of group dynamics and reasoning behind decision making—this is an extremely important reflective practice. Teacher-to-student questions can also be generative, critical, and reflective using a funnel approach, where they transition from general (divergent) questions, to focused probing questions (who, what, where, when, why and how), and then to confirmatory (yes or no) questions (Fig. 6.3).

6.4 Essentials Skills Developed in a Collaborative Makerspace

No one can whistle a symphony. It takes a whole orchestra to play it.

H.E. Luccock.

Collaboration as a support to metacognition

Students develop a range of higher-order thinking skills through the collaborate process including synthesis and evaluation of ideas and taking basic knowledge and applying it to new situations. When students become teachers or mentors to others when working on a collaborative project, they deepen their understanding about what they know. When students explain a concept to someone else, it invokes their own understanding. Vygotsky describes this as the **More Knowledgeable Other** (McLeod, 2018).

By highlighting the different learning styles of individuals within groups, students are provided the opportunity to see things more broadly than if they were just looking at the problem using their own conceptual schema. They begin to understand that there are many ways to view the world, to solve problems and generate new ideas. They expand their ways of knowing and their thinking toolkit of strategies and frameworks by sharing what they know and how they learn with others in the group.

Collaboration and interpersonal and intrapersonal skills

Affective and social emotional skills can also be developed through collaborative processes. These include resilience, perseverance, empathy, consideration of others, compromise, and patience. Described as Personal and Social Capabilities within the Australian Curriculum's General Capabilities, they are seen as essential for developing positive and powerful relationships between individuals. Makerspace activities provide a safe and objective context in which these skills can be developed. For example, if a group consensus is required it means that there is 'give and take' within the group as well as compromise to achieve a mutual goal.

Collaboration provides many rich opportunities for students to develop a range of communication skills including written and oral skills for a variety of purposes and audiences. These skills are further examined in Chap. 7.

How can teachers develop a collaborative classroom/Makerspace?

Teachers have a very important role in developing a collaborative culture in the classroom/Makerspace. Four main strategies include:

Model it

The saying 'they don't know what they don't know' is very applicable to a collaborative classroom where students may not have experienced or be aware of many of the expected behaviours. Therefore, it is important that the teacher models these behaviours, whether it be active listening, asking clarifying questions, admitting they were wrong, or negotiating outcomes. Teachers working together collaboratively as a team is also a powerful strategy that allows teachers to model the various skills they are trying to develop in their students.

Enable it

To develop collaborative practices in the classroom, students need to be given the opportunity to engage in those behaviours. There are many strategies that teachers can use to help develop collaborative behaviours at each stage of the BAR Process. See Table 6.2 for a comprehensive list of strategies and resources to assist with this process.

Monitor it

Monitoring the progress of collaboration needs to be actioned at multiple levels. Teachers need to ensure that individuals students, as well as groups, are on task to achieve their goals. Monitoring can take many forms and Table 6.2 also provides examples of strategies, templates and tools that will assist in the collection of information/data around the collaboration process. Monitoring is also part of the role of the team and needs to be supported and managed.

Celebrate It

It is important that not only successful project outcomes are recognised in Makerspace activities, but also desirable group traits and individual characteristics mentioned previously are celebrated. Who are 'great listeners'? Which group demonstrated active participation in the group norms? Who are great student mentors that help out others? These are all very important behaviours that need to be recognised and not just how successful the project was. An example of this can be seen at this Girls of Steel team building activity in Pennsylvania.

6.5 Operationalising the Collaboration BAR Model in the Classroom

Table 6.2 provides a comprehensive guide to ways that you can develop collaborative practices within your classroom/Makerspace and is derived from the BAR model in Table 6.1. It features the following:

- ***Skills/Strategies*** describes key aspects of the collaborative process that need to be considered/included;
- ***'What it sounds like'*** are examples of the types of questions you and your students need to be asking throughout the collaborative process; and
- ***'What it looks like'*** are ideas of templates and activities that will make the learning visible and provide evidence of the learning/process.

Table 6.2 Application of the collaboration BAR model in the classroom

Build	Skills/Strategies	What it sounds like	What it looks like
Process	• Identify the purpose, timeframe and desired outcomes of the collaboration • Determine how the collaboration will be documented/recorded • Identify steps to be undertaken throughout the collaboration • Allocate roles and responsibilities to group members • Establish communication processes and formats to be used	• What are the goals of this collaboration? What are we trying to achieve? (product, solution) • What is our timeframe for this collaboration? • How will we go about this task? Can we make is smaller and more manageable? • What are our success criteria? (How will we know if we have been successful or not?) • What are the milestones we are hoping to achieve? • How will we know if we have achieved them? • How will we document our progress? What tools, technologies and methods will we use to document our progress?	Create ***Project Management Process*** which will include: • SMART goals • Task and role allocations • Communication channels, formats • Timeline of actions • Project barometer (how are we going?) • Examples: • Design process • Kanban chart • Inquiry process or design thinking • Decision-making steps • Technology tools: • Trello Board • Google Docs (for synchronous sharing) • Visual learning templates and processes • Google classrooms • Microsoft Teams
Efficacy	• Establish group norms/team identity • Determine the purpose of the group/collaboration • Determine 'Rules of Engagement' • Students set goals based on their reflections on previous collaboration experiences • Individuals set personal goals for their engagement in the collaboration	• What should your group look like? • What will you call your team? • Who will be the leader in the group? • How will you choose what roles each person in the group has? • What roles will each group member have? • How will your group deal with conflict? • How will the group come to consensus? • How will you work together if someone has a different idea to you? • What will you do if you disagree with someone in your group?	• Roles and responsibilities template with task allocations made explicit • group contract • group norms articulated on group norm chart: • responsibility, communication, • active participation, conflict resolution, and respect. • Develop feedback process • *Feedback and Trust Grid* • 'Rules of Engagement' Contract (signed agreement of behaviours and expectations
ACT	Skills/Strategies	What it sounds like	What it looks like
Process	• Monitor group member actions to ensure roles are being adhered to • Monitor groups' ability to follow agreed collaborative process • Use agreed communication formats to record group actions	• How will you present your ideas? • Are you 'on track'? • What is working? • What isn't working? • What do you need to change? • What is stopping you from achieving your goals? • How are you progressing against the success criteria?	• SWOT analysis • PMI • Milestone review • Reflective journal • Checklist of actions • Use *Feedback and Trust Grid* • Use *Circler Review* as a check-in process to monitor progress

(continued)

Table 6.2 (continued)

ACT	Skills/Strategies	What it sounds like	What it looks like
Efficacy	• Monitor individual participation/active engagement in allocated roles • Monitor learners attempts to use alternative strategies when required • Monitor learners' ability to resolve differences according to the 'Rules of Engagement' • Monitor learners' ability to resolve differences according to the 'Rules of Engagement'		• Checklist of roles, actions to be completed, communication and problem-solving strategies used so far during course of collaboration • Use 'Rules of Engagement' guidelines to monitor progress
REVIEW	Skills/Strategies	What it sounds like	What it looks like
Process	• Evaluate the effectiveness of: • the outcome of the collaboration • the collaborative process • documentation processes • strategies used to document/disseminate the findings/outcomes of the collaboration	• What would you do differently next time? • Where might you get help from next time? • How effectively did we record the process of the collaboration? • How effective were our communication methods?	• Review all documentation used throughout the process • PMI chart • Think pair share • One minute papers • Four corners
Efficacy	• Evaluate the effectiveness of: • The established 'Rules of Engagement' • The allocation of roles and responsibilities of each group member	• What were the groups' strengths and weaknesses? • What problems did the team encounter, and how did they resolve them? • How do we know if the group was effective? • How well did I fulfil my role/contribute to the team? (individual)	• Review all documentation used throughout the process • Three stars and a wish • Yes/no cards • Opinion chart • 321 chart

Case Study—Pipeline Challenge Activity

The pipeline challenge is a STEM-based challenge with a strong engineering focus. Students must design and build a pipeline that has a set number of length and angle specifications. The students will have to design, construct, and test a paper pipeline that is able to transport a table tennis ball from one point of the classroom to another.

Materials—Ping-pong balls, A3 paper for drawings/design planning, A3/heavy paper for pipes, special angle joins and supports (five of each per group), sticky tape, scissors, tape measure or ruler, and protractor for measuring angles (Fig. 6.4).

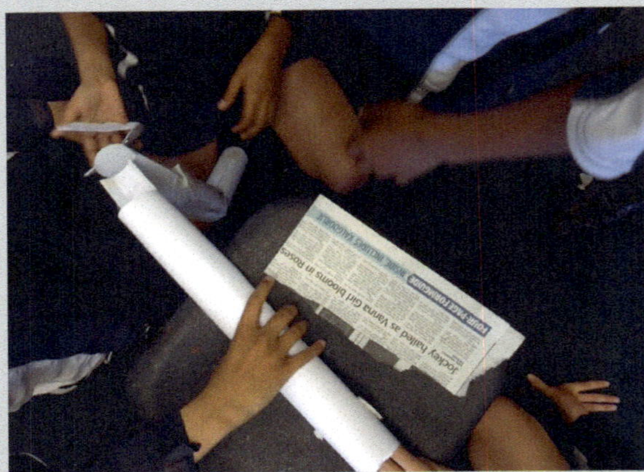

Fig. 6.4 Pipeline challenge

Design Brief

(i) The pipeline must be able to transport the table tennis ball without outside help.

(ii) The pipeline must have 3 angles, as the pipeline has to go around environmentally protected areas. One must be 90°. The other 2 must be at least 30°. Show each group the path they need to traverse in the classroom.

(iii) The starting point will be the chair height above ground and the finishing point will be at floor level, with necessary supports in between.

(iv) The only materials to be used are the ones provided by the teacher.

Discussing Design and Developing Each Other's Thinking

(Experimenting with moving the ball through the full pipeline).

Student 1 What if we make it very, very small so small so it is able to balance?

(The student is showing the others in the group where in the paper pipe he means).

Student 2 What if we make this drop here into the pipe will it be part of the pipe length? (*Student then takes over and bends the pipe at the point indicated).*

Student 3 (*interrupting*) You mean if we put it like this?

(*Showing them where to place the ball in the paper pipe*).

PST Let me check the rules so… (reading) 'the chair height will be the starting point', so have we started our pipeline in the right spot?

Students ask clarifying questions and then work together on the activity together all thinking and negotiating, with the preservice teacher working alongside them to ensure they meet the design brief. Through this and the students' dialogue, each student takes turns, and everyone attempts to problem solve the issue (Fig. 6.5).

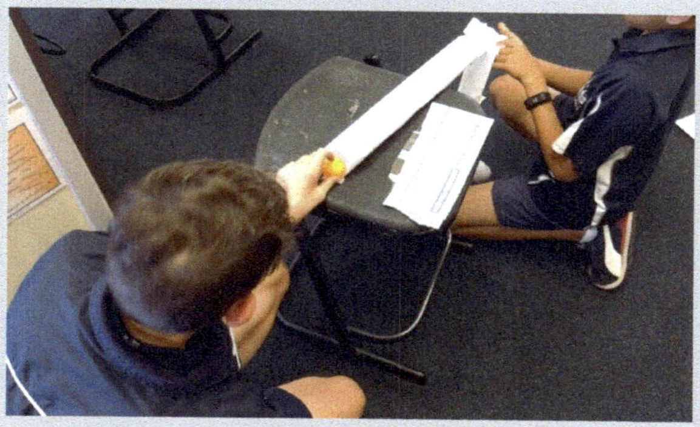

Fig. 6.5 Students and the pipeline challenge

Conclusion

Science concepts: gravity, fluid flow, friction, strength of materials, timing, pressure, bouncing/reflection.

Maths concepts: surface area (to calculate the material needed for the pipeline), measurement.

Engineering concepts: design, design changes.

Coming together is a beginning, staying together is progress, and working together is success.
Henry Ford.

6.6 Summary

What is collaboration?
Collaboration occurs when two or more people work together to create or achieve a mutual goal. In schools, collaborative learning is the use of small groups where students work together to maximise their own and each other's learning.

How can Makerspaces support collaboration?
Makerspaces provide opportunities for individuals to work and engage with others to create a product or solve a problem. Even when students are working on an individual project, they will often collaborate (both directly and indirectly) with others, either mentoring and providing expertise, or seeking support from a more knowledgeable other. This can occur in both a formalised Makerspace with a well-developed collaborative structure such as groupings/group norms, or a less rigid Makerspace environment where collaboration occurs organically.

What is the BAR model and how can it help teachers create a collaborative space?
The BAR model identifies three phases of the dynamic collaborative process: Build, Act, and Review. Process and efficacy are defined as key components.

What can teachers do to promote collaboration?
Teachers are encouraged to focus on the following to promote collaboration in their Makerspace:

- *Model it* by working in a team with colleagues;
- *Enable it* through the strategies and resources in the BAR model Table 6.2;
- *Monitor it* through the strategies and resources in the BAR model Table 6.2;
- *Celebrate it* through the acknowledgement of individual characteristics.

Chapter 7
Developing Communication Skills in a Makerspace

Communication sometimes is not what you first hear - listen not just to the words but listen for the reason (Catherine Pulsifer)

Keywords Communication · Collaboration · Information · Synchronous · Asynchronous · Transmissive

Definition of Communication

Noun The imparting or exchanging of information by speaking, writing, or using some other medium.

Verb To share or exchange information, news, ideas, and feelings (Oxford Dictionary).

Focus Questions

- What does communication look like in a Makerspace?
- How can we create different types of purposeful communication in a Makerspace?
- How can educators create questions to support a range of learning outcomes in a Makerspace?
- How do we capture evidence of communication in a Makerspace?

R. B. Koul et al., *Teaching 21st Century Skills*,
https://doi.org/10.1007/978-981-16-4361-3_7

7.1 Introduction

This chapter discusses the importance of communication development in the context of a Makerspace. By adapting specific strategies in the classroom, teachers can encourage students to strengthen their communication development. The efficacy of these strategies can significantly impact a student's written and verbal communication skills both in and out of the classroom. After delving deeply into the forms of communication, the centre of this chapter considers communication in a Makerspace as the imaginative and creative space to develop each child's narrative through a diverse range of tools, technologies, and artefacts, which departs from the expected positions of reading and writing or sharing ideas.

7.2 Defining Communication

Communication is a two-way process. It takes place when one person transmits information and understanding to another person. It can occur synchronously (real-time) or asynchronously (static). Rock paintings (Fig. 7.1) may be considered one of the first forms of communication that has lasted millennia. Since the dark ages, the forms and mediums of communication have evolved and become more complex, the essence of what communication is used for remains the same. To explore how the purpose, forms, and mediums within a Makerspace learning environment facilitate communication skills development, educators must first be able to understand broad communication concepts.

7.3 The Communication Process

Figure 7.2 shows how communication can be considered a reciprocal process between two or more people. A communicator sends a message through one of many communication channels to relay a message. Once this message has been received, the audience can then communicate (respond) via feedback. For example, one of the most simplistic types of communication is listening and responding to another who is sharing information. The movement of lips, the wave of hands, or a swift glance may convey as much (or sometimes more) meaning as written or spoken words. The basic elements of communication in this model are the communicator, audience, message (which includes consideration of the purpose), channel/medium used, and response. Receiving includes both how we take in the message (reading, listening, or viewing), and the decoding of the message. This decoding allows the receiver to interpret the meaning of what has been communicated. Each of these elements is defined as follows:

Fig. 7.1 Communicating through rock paintings

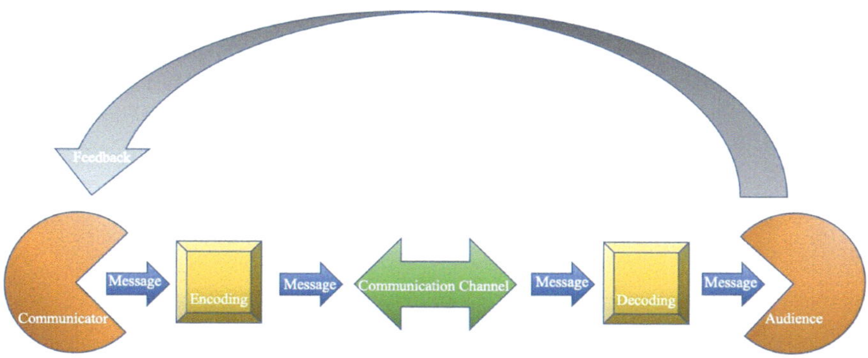

Fig. 7.2 Communication process

The **Communicator** (sender, speaker, writer, etc.) expresses or transmits a message. They have developed an idea about concepts, beliefs, or data that they want to convey. An appropriate and relevant medium for transmitting this message is then selected. Key aspects used to reinforce or strengthen communicators ability to exchange information include:

- Physical: How a communicator uses their body language, facial expressions, and voice;
- Linguistic: The communicator's use of language, including their understanding of formality and rhetorical devices, as well as tone, expression, and inflection;
- Cognitive: The concepts, ideas, or information that the communicator wishes to share and their ability to build on, challenge, question, and summarise others' ideas; and
- Social and emotional: How well a communicator listens, includes others, empathises, and responds to their audience.

Encoding relates to intricate linguistic choices or expressions applied to the medium that allows it to be further understood by the audience. This includes things like grammar, syntax, semantics, and language conventions. Encoding applies to written, verbal, or symbolic (pictorial) communication.

The **Communication channel** is the medium through which the sender passes the information to the receiver. This channel can be wide and varied and may include visual, audio, sensory methods. Typical of twenty-first-century communication, a number of these channels will also contain technology.

Decoding The person(s) who receives the message or symbol from the communicator attempts to convert the information into a format that allows for meaning to be established. This must occur for the receiver to extract meaning to his/her complete understanding.

The **Audience** or *receiver* of the message for whom the communication is meant. The audience receives the information.

The **Message** is the subject matter of this process, or what is being communicated. This includes the content of the speech, order, words, views, feelings, attitudes, symbols, pictures, music, information, ideas, or suggestions.

Feedback How does the sender or communicator know that the audience has understood the message? Is it non-verbal, verbal feedback/restating an idea/asking clarifying questions?

> We take communication for granted because we do it so frequently, but it's actually a complex process.
>
> Joseph Sommerville (2016).

Communication and Maslow's Hierarchy of Needs

The capacity for self-expression and ability to communicate thoughts, feelings, and ideas competently is essential at all levels of Maslow's needs. Healthy and effective communication skills can ensure basics needs are met: seeking help when feeling unsafe, engaging in positive social discourse, and reflecting on one's own thoughts and feelings. Within a Makerspace environment, children will be actively involved in sharing ideas in a group setting when solving a problem, explaining why they think a particular process is the right course of action, asking questions, answering questions, recording their findings, and sharing what they have learnt. The activities in this chapter support the **Belonging, Respect,** and **Self-Actualisation** levels of the hierarchy (Fig. 7.3).

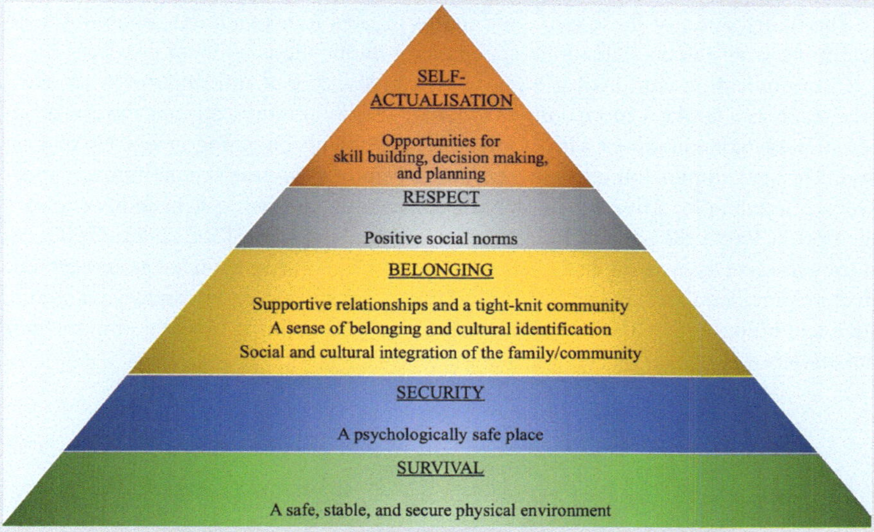

Fig. 7.3 Maslow's hierarchy of needs

7.4 Characteristics of Communication

Communication is a dynamic process with multiple elements acting simultaneously. While many of the contemporary models and definitions of communication compartmentalise these elements, they are symbiotic in nature and difficult to separate and define. One of the more relevant models that consider the nebulous nature of communication is the communication quadrant model featured in Fig. 7.4. This model identifies two key characteristics of communication and represents each of them on a continuum:

1. Transmissive versus collaborative communication; and
2. Synchronous versus asynchronous communication.

The intersection of these two continuums creates four quadrants, enabling us to determine where in each quadrant specific communications skills or concepts lie.

Communication can involve a one-way *transmission* of information to an audience, such as a book. A more *collaborative* mode of communication is the construction of new understandings with others, such as students engaging in a group discussion. These examples fall into the *transmissive* and *collaborative* quadrants, respectively. For example, a live TV news bulletin is a transmissive, synchronous communication, whereas the ancient Indigenous rock art at the start of the chapter is a form of transmissive asynchronous communication. Collaborative asynchronous communications include the use of shared online team workspaces, email, and social media. Such technologies can also be used synchronously, depending on the communicator and audience.

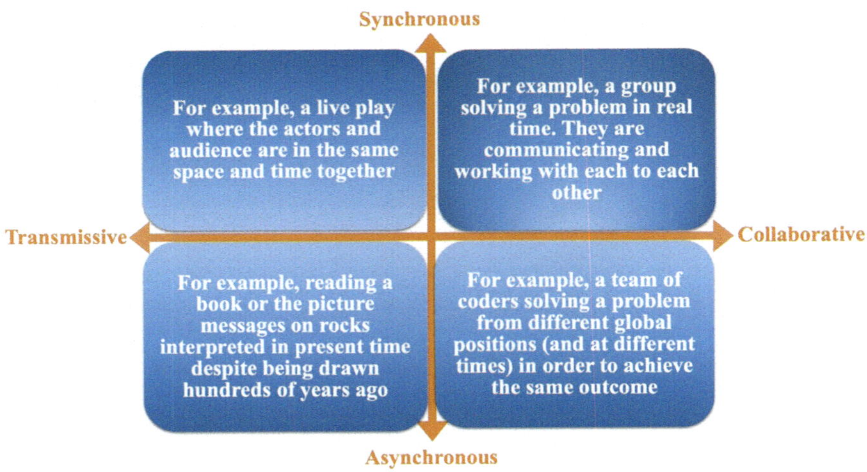

Fig. 7.4 Communication quadrant model

7.5 How Makerspaces Can Help Develop Effective Communication in STEM

> When I travelled as an engineer, I'd fly to Jakarta and we'd jump off the plane and we'd have to deal with a different language, a different culture, a different set of norms, a whole different history, and we had to be cognitive of all those things. Then we'd jump in a plane and fly to Japan and then to Thailand and India and Singapore. It wasn't your ability to be technically competent that allowed you to solve the problem, it was your capacity to communicate your idea.
>
> Meyes (2020 Re-Engineering Australia Foundation).

Makerspaces provide opportunities for students to develop a repertoire of communication skills as they are a more open, inclusive, and flexible learning spaces. Through speaking, listening, reading, and writing, students can facilitate their own expression of ideas and questioning to develop these key communication skills.

The most common form of collaborative communication between students is speaking and listening as they work. They discuss ideas, make generalisations, justify arguments, and challenge the thinking of others. The brainstorming and ideation processes of a project or task are also a crucial step. The development of active listening ensures that students understand what has been said by others in a group setting. This process also allows students to articulate their own thinking through talking.

Student writing can take on many forms, whether it is creating a Mind Map, using a graphic organiser to structure ideas, writing up a process or design plan, or reflecting on their work. Developing strong writing skills benefits students when they are required to present ideas or findings. Collaborative communication is a skill that is developed organically as students work within a Makerspace environment, as open and collaborative communication is one of its main concepts.

7.6 A Teacher's Role in Developing Communication Skills

It is the responsibility of teachers to create opportunities for communication skills within their Makerspace to be strategically developed through questioning, scaffolded processes, and activities. Many of these skills must be strategically taught, modelled, and practised, ensuring they are internalised for long-term development. Teacher behaviours for improving student communication include:

- the establishment of group norms and structure for discussion/communication;
- assisting students to determine their roles in the communication process which involves assisting students to generate effective, open questions to guide their inquiry and problem-solving processes;
- developing etiquette for the chosen communication format (e.g. raising your hands, no phones in class, using a salutation in an email, etc.);

- providing multiple opportunities for children to communicate in different contexts using different mediums; and
- creating a culture of trust and openness.

Students need to be able to communicate what they know, or what think they know. The best way for teachers to facilitate this from all students is through classroom discussion and small group work (Rika, 1996). Students as mentors is also a powerful communication strategy as students must verbalise their own knowledge and understanding of a topic to others. They are able to benefit from hearing concepts being explained from different points of view and also in ways that might be closer to their own way of thinking.

'Think time' is a strategy that provides students with the opportunity to consider their response to a question without pressure. This allows students time to think about the most effective way to communicate to their teacher, their classmates, and themselves. Open-ended questions and discussion promote higher-quality communication rather than a simple 'yes'/'no' dialogue. Teachers should avoid these yes/no and short answer questions if they want to facilitate quality discussion. Open-ended higher-level thinking questions are the best choice to encourage students to think more deeply and effectively communicate their ideas. A teacher's involvement in a discussion should also be sensitive when correcting misinformation. They need to constantly model the process of active listening with empathy. Such attributes exhibited by a teacher enable the creation and maintenance of a rich environment where students feel comfortable to participate.

7.7 Teacher Questions to Support Students Communication

The purpose and nature of communication changes with every interaction. Wilber (2020) dissects the ongoing communication process into three essential stages: building, action, and review. Many of the proposed questions featured in Table 7.1 below are better suited at the beginning of a communication process (build), while others are most effective within the 'act' and 'review' sections. The 'Build, Act, Review' cycle demonstrates how the communication process can be ongoing.

7.8 Case Study

Using Questions as a Form of Communication

Students in this Grade 6 Makerspace classroom were provided with materials and a Wigglebot model and challenged to create their own working model from the pieces. Interviewers were encouraged to question them about what they were doing and not tell them the answers. This Makerspace project was intentionally focused on determining students understanding of the content

Table 7.1 BAR process and questioning

Build	Act	Review
Process	*Process*	*Process*
Identify the purpose, timeframe, and desired outcomes of the communication type (presentation, essay, etc.) Establish communication processes and formats to be used	Monitor group member actions to ensure roles are being adhered to Monitor groups' ability to follow agreed collaborative process Use agreed communication formats to record group actions	Evaluate the effectiveness of: • the outcome of the collaboration • the collaborative processes • the documentation processes • the communication strategies used to disseminate the findings / outcomes of the collaboration
Questions	*Questions*	*Questions*
What is the purpose of your message? Who is the audience of your message? How will you deliver your message to the audience? How will you make your message interesting to the audience)? How are you presenting to the logic, feelings, and credibility of your audience?	How can you pitch your idea/message in a clear and logical way? In your conclusion, restate your main point and identify the action What is a suitable example? What is a suitable quote?	What is your central idea? How have you clearly stated your central idea (usually at the beginning)? Is your language clear, specific, accurate, and appropriate to the audience, purpose, and medium? Is your language clear, specific, accurate, unassuming with no clichés and misused jargon? How have you explained technical language and terms? If so, is it clearly defined and explained (depending on the knowledge of the audience)?

knowledge (in this example, science content knowledge) to determine if there are alternate conceptions, misconceptions, or misunderstanding of the project and how this affects a students' ability to complete a task.

This case study focuses on effective communication strategies between teachers and students when involved in a Makerspace environment (Fig. 7.5).

Fig. 7.5 Wigglebot example

Student
So we have connected the battery to the motor and connected the peg to the motor. When the battery is on, the electrons flow through to the motor and then the motor will spin making the peg move.

Teacher—Question
Why are there two wires, do you need two wires, or do you just need one?

Student
You need two, as one gives the positive and one gives the negative (Fig. 7.6).

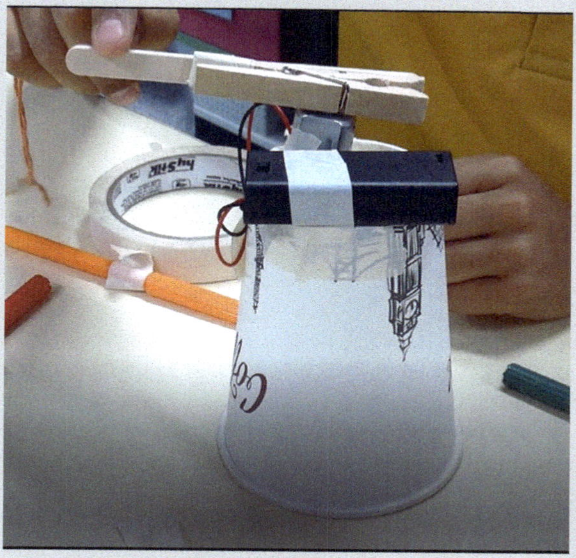

Fig. 7.6 Wigglebot construction

Teacher—Question
What do you mean, can you explain that? What does that have to do with the electrons?

Student
There are positive and negative electrons. The battery gives off positive electrons, well, one gives off positive and one gives off negative. So, both of them make the motor spin.

Teacher—Question
What is the peg for?

Students
The peg is so that when it spins then it will actually wobble. So, since this side will weigh more (with pop stick), then it will shake a bit.

Teacher—Question
How are you going to put the legs in?

Student
Split it into thirds and stick it on each third.

Teacher—Question
Why each third?

Student
Because thirds then it would equal. One side wouldn't be heavier than the other.

Teacher—Question
How will you work out where a third is?

Student
I can estimate, and if that does not work then I'll use a bit of trial and error.

The second piece of evidence in the same classroom Makerspace was a short survey to probe the students understanding. These were written questions but were open and focused on eliciting content knowledge. Here we can see that the student is confident that they can identify the science and then discusses the most interesting aspect of the activity and complete a labelled diagram. These students in this Makerspace had had science classes that had taught them the concepts of electricity and energy transformation but it was clear from other student interviews that some were not confident and had some misconceptions in this area (Fig. 7.7).

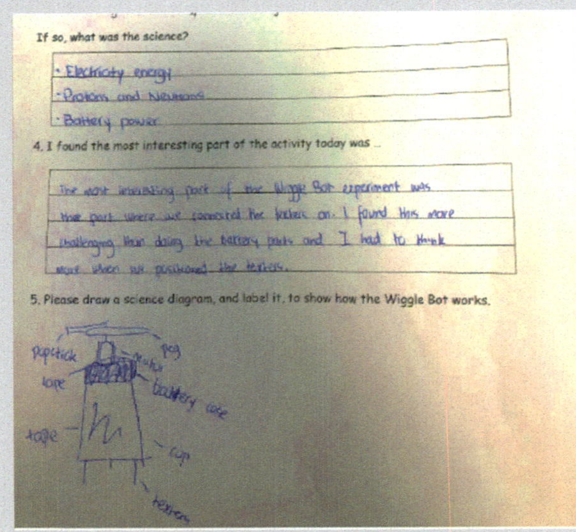

Fig. 7.7 Reflection questions

The student can list the science as three words 'electrical energy', 'protons and neutrons', and 'battery power'. But there is no other explanation, and so in this communication, it is difficult to know what understanding that the student has. The teacher may interpret here that the student does not understand electricity deeply as they are talking about protons and neutrons and it is the electrons that are involved in the process of electrical energy.

7.9 Beyond Regular Communication

Communication in a Makerspace is not just reading and writing, literacy or numeracy, and communication between individuals. While these are all incredibly important, they are covered more frequently in the classroom. Students must develop these skills to be able to read through their textbooks and written materials and frequently express their understanding through writing. Makerspaces are not devoid of these aspects: there are a vast array of wall posters and instructions that students must read, while note-making to remember and communicate engages their writing skills. For example, exploring the opportunities that students have to write in code and create a robot to help them develop their STEM understanding bridges an array of affective Transversal Competencies discussed in Chap. 5 (Fig. 7.8).

Makerspaces can be an exciting space for students to demonstrate their learning and understandings in STEM governed by much broader parameters than those usually set in regular classrooms. Students can use an incredibly large array of tools and technologies to demonstrate their understanding of their learning, and this provides those students who do not experience success due to issues in writing

Fig. 7.8 Laser cutter

and reading to blossom and thrive, often surprising themselves and their teachers in the process.

Makerspaces also provide students the opportunity to expand on typical modalities of verbal (presentation or conversation) and written (essay or report) communication. In the same way, a photograph or piece of art is a form of communication, so too is the range of other tactile representations available in a Makerspace. This includes the creation of artefacts with tools such as cardboard, laser cutters, and 3D printers which are outlets to develop both collaborative (construction of a group project, for example) and transmissive communication in a twenty-first-century classroom. The examples below feature various ways teachers can nurture communication development while students can express and adapt to a more hands-on approach to communication (Figs. 7.9 and 7.10).

In the following table, we list and provide examples of student communication in a Makerspace considering the types of message and how it can be represented (Table 7.2).

Fig. 7.9 Example of a Makerspace project. Image from https://woodguide.org/guide/laser-cutter

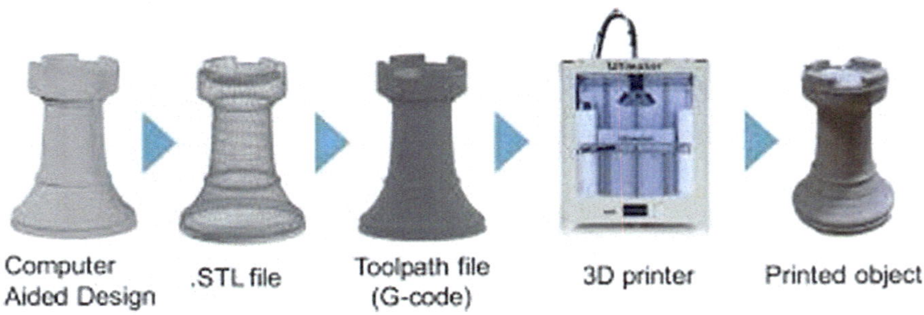

Fig. 7.10 3D printing process chain

Table 7.2 Communication domains and corresponding examples

Domain	Examples	Example of tools
Speaking	Podcast	Podcasting apps Audacity (Win/Mac) GarageBand QuickVoice App Voice Memo App
	Oral presentation	Live as a face to face Virtual using WebEx, Skype, Google Meets,Zoom
	Discussion (Chap. 6 Collaboration)	Students planning group task either face-to-face or through online collaborative environment, e.g. Minecraft community (Chap. 3, see example Bee Hotel in Minecraft), Trello Board, Google Docs
Reading	Written instructions	Using written instructions to understand a tool
	Programming language	Reading visual programming languages such as Scratch, Scratch Junior, Microsoft MakeCode, and Tynker Reading text-based programming languages such as HTML or Python Code Club, CoderDojo, Hour of Code, Code Academy, and Google CS FIRST offer programming curricula to support programming clubs in schools)
	Iconography	Ability to interpret tool symbols in Minecraft, 3D printing software
Writing	Documenting their progress	Scientific report PMI Reflective journal Engineering Design Log/Notebook
	Programming a robot	Writing programming languages such as Python, Scratch, or HTML
	Creating a mind map	Digital - Inspiration, Bubbl.us, and Popplet Non-digital - using some paper and a pen
Artefact	Physical computing and robotics	Use of coding languages to control and program physical computing devices such as BBC Micro:bit, MakeyMakey, Arduino, and Raspberry Pi Use of coding languages and tools (where applicable) to construct and control robots and electronic sensors, e.g. LEGO Mindstorms, Spike Prime, Beebots, Dash and Dot, Sphero, MakeBlock mBot

(continued)

Table 7.2 (continued)

Domain	Examples	Example of tools
	3D Printing	This can include the entire process of developing the 3D item from the beginning using a modelling program such as TinkerCAD, Google Sketchup, or Blender; or printing a 3D file downloaded from Thingiverse, Instructables, Maker's Empire, or YOBI 3D
	Constructables	For example, Wigglebot (See Case Studies) that can be constructed from materials that are presented to the student or that the student sources Clay, recycled materials, cardboard, fabric, and other textiles can be used here
Multimedia	Animation	For example, Claymation, PuppetPals HD, Powtoon, Pixton, Flux time, LEGO Stop Motion
	Infographics	Canva, Visual.ly, Easel.ly
	Movie	Animoto, iMovie, Windows Moviemaker, Apple Clips, WeVideo, Greenscreen by DoInk, Sock Puppets, and Chatterpix
	eBook	Using an iBook Author or Book Creator
	Website /blog	Google Sites, Blogger, Edublogs, and the use of site templates from Weebly, WIX, and WordPress

7.10 Summary

While we have discussed the traditional forms of communication and their role within a Makerspace, we acknowledge the invaluable role of Makerspace in developing opportunities for students to communicate their understandings, maturation, and beliefs that are often not evident within traditional domains.

What does communication look like in a Makerspace?
Communication within a Makerspace is complex and nuanced and can be driven by teacher's observations of body language and listening to students' communications. This allows teachers to determine how concepts are individually learnt and how skills are expressed. Communications are diverse and can often move away from the more traditional forms of messaging and can be seen, for example, in the usefulness of a 3D artefact designed and printed by the student.

How can we create different types of purposeful communication in a Makerspace?

Once you have considered the nuanced nature of Communication and how human communities are perpetually communicating on all levels, it is important for educators to focus on giving students a wide range of opportunities to communicate via teacher questioning, group dynamics, reflective processes, and different projects with varied tools and materials.

How can teachers create questions to support a range of learning outcomes in a Makerspace?

Questions are pivotal to help students demonstrate their understanding and communicate their feelings. While there are no set questions the use of questions is often a powerful way of starting to collect evidence of learning, skills, and feelings.

- Who;
- What;
- Where;
- When;
- Why; and
- How.

How do we capture evidence of Communication in a Makerspace?

Evidence of communication can be collected through a vast rage of methods, and many are included in Table 7.1, which considers the types of tools that teachers can use to help students communicate their message. These are wide and varied and move beyond the important (but often over-focused) reports. These tools also encourage an imaginative way for the teacher to support communication.

Chapter 8
Developing Resilience in a Makerspace

It is not easy to teach resilience in the classroom - but it is crucial (Tocino-Smith, 2019)

Keywords Resilience · Perseverance · Grit · Social competence · Positive self-image

Definition of Resilience

The capacity to persevere through a challenge or recover quickly from an adverse event, or period of adversity.

Focus Questions

- What is resilience?
- Why is it important to develop resilience in children?
- How can a makespace help develop resilience?
- What does resilience look like in a Makerspace?
- How can resilience be assessed?

R. B. Koul et al., *Teaching 21st Century Skills*,
https://doi.org/10.1007/978-981-16-4361-3_8

8.1 Introduction

Angela Duckworth popularised the term *resilience* in her 2017 book *Grit: The power of passion and perseverance.* She defines resilience as an 'indispensable trait needed by individuals to live in society today'. Resources she created such as the 'Grit Scale' and 'Character Lab' attempt to deconstruct this nebulous term into identifiable behaviours and traits by suggesting ways they can be developed within a classroom. Her book was also a catalyst for research around the importance of building resilience in students while they are at school to prepare them for their future. By defining resilience within the context of Makerspace environment, it is possible to identify learning strategies to both develop and observe resilience in students.

> When students believe they are worthy and capable of overcoming challenges, they become resilient.
>
> Juliette Tocino-Smith (2019)

In order to understand resilience, it is crucial to identify what it *is* and what it *isn't*. According to Beyond Blue's 2017 study, *Building resilience in children 0–12: A practice guide,* resilience:

- develops through individual and environmental factors and how these two aspects interact;
- falls on a *continuum;*
- *evolves over time;*
- can be *learned* (given the opportunity and experience);
- will *manifest differently* depending on an individual's environment and upbringing.

Resilience is not

- a fixed character trait;
- only focused on the skills and competencies of the individual;
- something we are born with;
- an absence of negative emotions.

Resilience is not a single behaviour that can be developed through one experience or identified in a single observation: its development and identification are facilitated over a period of time through multiple experiences.

> Resilience is something that every child can learn, and it's important that every child is provided with opportunities to build resilience.
>
> Beyond Blue (2017)

8.2 Resilience and Maslow's Hierarchy of Needs

'Resilience is about the process of becoming, which children understand once they develop a firm belief about their place in the world' (Tocino-Smith, 2019). This concept is at the very heart of Maslow's Hierarchy of Needs discussed in detail in Chap. 1. If we consider the apex of this original hierarchy, self-actualisation, as seeking and realising personal potential, self-fulfilment, and ultimate personal experiences, then developing resilience in a student is an essential disposition needed to achieve this goal. Students will develop resilience when they have:

- a secure and safe physical environment (**Survival**);
- a psychologically safe place where they are encouraged to take risks (**Security**);
- Feelings of belonging and identification where they have clear roles and responsibilities in the group (**Belonging**);
- social and cultural integration of the family and community (**Belonging**);
- supportive relationships in a Maker community (**Belonging**);
- positive social norms (**Respect**);
- opportunities for skill-building, decision-making, and planning (**Self-actualisation**).

Makerspaces can provide all of these elements through their programs and activities. The very nature of a Makerspace as a 'safe place to come and try stuff out' encourages students to explore, experiment, fail, and try again without judgement, and in that exploration, they push themselves further than they might in a traditional classroom setting. Using Maslow's hierarchy as a guide when developing the space and program for a Makerspace allows this to be achieved in a strategic way (Fig. 8.1).

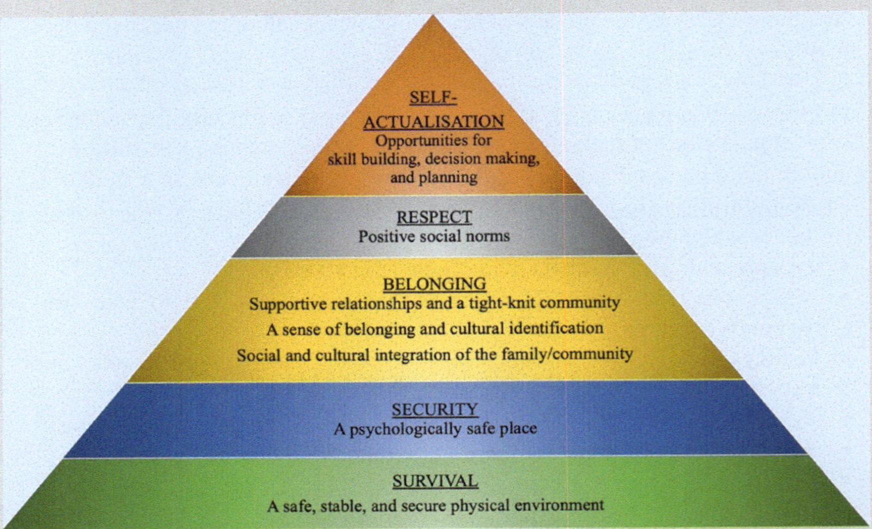

Fig. 8.1 Maslow's hierarchy of needs

8.3 Language of Resilience

In the past quarter century, scholars, psychologists, and educators have all attempted to define resilience and, while terminology may vary, there appear to be common characteristics evident in the resilient individual. These characteristics include commitment; recognition of limits to control; engaging the support of others (seeking help); nurturing close, secure attachment to others; establishing personal or collective goals; self-efficacy; developing a realistic sense of control and sense of humour; having an action-oriented approach to problems; and having patience, adaptability to change, optimism and faith (Connor & Davidson, 2003). Cahill and Beadle (2014), Farrelly, Forster, & Smith (2014), and Ginsberg (2014) determined that the following factors also contributed to the degree to which an individual can be considered resilient. These include social competence, optimism, purpose, coping strategies, positive self-image, confidence, connection, control, efficacy, and agency. When students experience self-efficacy, they experience the feeling of being in control of the world around them. In this way, they learn to accept consequences for actions they have taken, both positive and negative. 'Owning their actions' is an important stage in developing emotional maturity. This empowerment gives them permission to make choices and take risks.

> Children who develop resilience are better able to face disappointment, learn from failure, cope with loss and adapt to change. We recognise resilience in children when we observe their determination, grit, and perseverance to tackle a problem and cope with emotional challenges of school and life.
>
> Marilyn Price-Mitchell (2015)

8.4 The Role of Teachers in Developing Resilience in a Makerspace

> When educators believe in their students and prioritise growth, students are able to reflect on their own progress and develop resilience.
>
> Namka (2014)

Once an educator is able to identify the characteristics of resilience, they must then consider how to create opportunities for students to develop these behavioural traits in a Makerspace. For example, how do we develop *coping strategies*? Is this through visual elements like posters and displays, or through class rules and *group norms*? Can *a sense of purpose* be nurtured through the development of personal and group goals? How do we create a risk-taking culture that allows children to *experience failure* in a non-threatening way? When creating a Makerspace, engineering these characteristics into the program will help foster a resilience-conscious environment. Classroom dynamics and teaching methods are vital tools when developing a classroom culture of resilience. Benoit (2013) highlights the importance of a teacher's function in resilience development by suggesting one of their fundamental duties as instilling resilience behaviours in their students. Therefore, an important role of teachers is to ensure students are encouraged to become risk-takers and are treated in a way that indicates to them that they can be trusted to think for themselves and act to test hypotheses and make decisions. Rosin (2014) reinforces this valuable role by suggesting that children need to be in the company of adults that they trust and who trust them in return; otherwise, they do not build 'the confidence to be truly independent and self-reliant' (2014). Ginsberg (2014) suggests that 'young people live up or down to [the] expectations we set for them…. [and] need adults who believe in them unconditionally and hold them to the high expectations of being compassionate, generous, and creative'. It is imperative, therefore, that teachers not only model resilient behaviours in the classroom but also regularly validate students as independent learners and valuable class members.

The Kintsugi Metaphor

The notion of resilience can be seen in the traditional Japanese art form of *Kintsugi.* Kintsugi translates to 'join with gold' and is an artistic practice whereby broken pieces of an accidentally smashed pot are reassembled to create a new object. The rejoining process features gold paint or glue to emphasise the rebuilding process.

The gold cracks in the finished product celebrate the imperfections of the bowl rather than trying to hide them. Think about how this metaphor could be used to help develop resilience in a Makerspace context.

> It is important to remember that the objective is not about encouraging success or perfection. The purpose is to value the attempt over the outcome and challenge over victory.
>
> Carol Dwek (2015)

Building children's resilience is a complex process that requires a multifaceted approach. It requires enhancing their environment while simultaneously developing resilient capabilities and behaviours. We start with the notion of Head, Heart, Hand, Relationships (Community) and then three themes grouped under Reflection. The strategies listed are designed to encourage teachers to think about being more resilience-minded while also building resilience into their own teaching styles. Resilient thinking, reflection, managing emotions, and building on strengths are cornerstones for resilience and are included in the domains outlined below:

Resilient Thinking (HEAD)

- Make resilience transparent / visible by talking about it. Ask students probing questions that require them to think about resilience strategies they can employ to solve problems.
- Provide children with opportunities to solve a range of problems both teacher- and student-directed.
- Encourage children in a positive way when they are undertaking a challenging task or activity.
- Suggest / model alternative strategies to overcome problems put forward.
- Discuss how students can plan to overcome difficult situations.
- Encourage academic self-determination and autonomy by allowing students to set their own goals.
- Use open-ended questioning.
- Promote a 'victor' not 'victim' mentality where children use a positive mindset to reframe situations.

Effective Relationships (RELATIONSHIPS)

- Provide opportunities for children to develop positive relationships with peers, teachers, and others.
- Encourage children to practice and improve social skills. Introduce / use a range of collaborative group skills.
- Discuss in groups how children can help others facing difficulties or hardship.
- Nurture effective peer relationships by encouraging students to work together as teams so as to develop problem-solving and conflict resolution strategies.
- Help students recognise and appreciate each other's positive attributes to enhance relationships.
- Collaborate with students to develop a group agreement that sets the parameters for behaviours in Makerspace. Give students the opportunity to contribute to these established group norms.

Managing Emotions (HEART)

- Provide children with opportunities to practice empathy. Use strategies such as Other People's Viewpoint (OPV) and Strengths, Weaknesses, Opportunities, and Threats (SWOT Analysis).
- Develop healthy thinking habits to help children regulate their emotions such as self-reflection, an optimistic mindset, positive self-reassurance, and an affirmative attitude.
- Validate a child's feelings by acknowledging and validating their feelings or concerns.
- Create environments that promote children's sense of belonging.
- Help children express their feelings by putting words to their emotions. Use tools like emotion wordlists, word clouds, and feelings ratings.

Building on Strengths (HAND)

- Provide students with opportunities to participate in healthy risk-taking activities.
- Explain the benefits of facing challenges. They can help us develop emotional maturity and grow as an individual. These skills and knowledge better prepare us for future challenges.
- Talk to students about preparing for future situations or events they may be feeling anxious about.

- Reaffirm that asking for help and support is OK. Brainstorm with children about who the best places and people to seek help and advice from (counsellors, teachers, parents, friends, etc.).
- Give children opportunities to experience everyday adversity. Set up some fun challenge activities.
- Build self-efficacy by programming academic instruction that is challenging and consciously support students in mastering those challenges.
- Incorporate mindfulness, breathing activities, or other relaxation techniques into everyday activities (Beyond Blue, 2017).
- Encourage behavioural self-control by creating a classroom climate where students are encouraged to think ahead and evaluate their own behaviour and how they respond their engagement in their learning.

8.5 Reflective Strategies

Mastery versus 'Not Yet'
So much emphasis is placed on the 'right' and 'wrong' answer that the idea of making a mistake or not understanding tasks are seen in a negative way. Students need to understand that to become a master or expert at something may take hundreds of attempts.

Students need to be introduced to this concept of 'mastery' and 'not yet' as a journey where practice makes progress. In a Makerspace, it may be an opportunity for you, the teacher, to profess that you do not have a mastery of every tool and process, and that you are prepared to show them how you approach learning and what happens when you do not complete a task perfectly. Encouraging students to 'make an attempt' and verbalising the idea of 'not yet' can be powerful strategies when used in a careful and meaningful way.

Reframing Failure
Just as Thomas Edison looked at failure as a necessary part of success, so too should students be encouraged to reframe the way they see failure—not as something to be avoided but as an essential stage in the pursuit of skill mastery. Learning cannot take place without failure.

Students should also be aware that when faced with adversity or a challenge, others are there to help them achieve their goals. Makerspaces should be areas of low stakes in a school, that is, where 'failing to achieve' will not preclude students from further study avenues or post them at the bottom of the class. Creating a school and classroom where every task is high stakes make students anxious and risk averse. Makerspaces can be joyful spaces to try out ideas and creations.

Attempt over Outcome

The phrase 'it's the journey not the destination' can be applied in a Makerspace where students are encouraged to focus on the process, rather than the end result. Too often important learning opportunities are missed because the only thing being assessed is a finished product (model, report, presentation, etc.). If this does not meet the necessary criteria, it is considered a 'fail' without assessing the process against the final product. By acknowledging the work completed throughout the process and not just the end result or desired goal, students are able to have a more holistic view of their learning and the skills they have used. This is why it is as important to assess the process as well as the finished product. Any attempt a student makes, regardless of outcome, should be acknowledged and seen as an opportunity for learning.

8.6 Developing Resilience Through Effective Questioning

Questioning is a powerful strategy that teachers can use to encourage students to reflect on their thoughts and feelings and articulate strategies they used to deal with difficult situations or problems. Verbalising abstract ideas and concepts can help students not only gain a greater self-awareness, but it also encourages them to 'talk their walk'. Questioning to develop resilience can be done in any and every stage of a lesson or activity. They can be explicitly or implicitly stated. They can be asked about what was 'thought', 'done', and 'felt' by each child. Some of the question's teachers may ask students include:

- What were some of the challenges you came up against while doing this activity?
- What did you do when you came up against a problem?
- How did you feel when you came across a problem?
- How did you feel when you made a mistake?
- What strategies did you use that helped you solve the problem?
- When did you decide that you were not going to be able to solve the problem using your chosen strategy?
- Where might you get help from next time?
- What would you do differently next time?
- When did you decide that you had to completely change direction / abandon the project?
- How did that make you feel?

When observing students in a Makerspace, whether working on their own or in groups, there are conversations occurring both formally and informally. Student self-talk and group discussions all provide rich evidence of resilience.

8.7 Assessment—Measuring Resilience

The very nature of resilience as an interconnected set of behavioural traits makes it very difficult to measure, but not impossible. Resilience can appear significantly different in each and every child in your class and a child may also demonstrate varying degrees of resilience over time depending on what is currently happening in their lives. This means the question of how to measure resilience is very difficult to answer.

While there is no single skill that encompasses the trait of resilience, there are a number of resilience scales that attempt to quantify it. The Brief Resilience Scale (BRS) is an effective way in measuring and assessing resilience, and the ability to recover reliable means of assessing resilience as the ability to bounce back or recover from stress. The Connor-Davidson Resilience Scale (CD-RISC) is also an accurate tool in distinguishing resilience development in students.

Beyond Blue (2017) also suggests that resilience should be measured consistently over several points in time, where possible both qualitative and quantitative data should be collected. Quantitative data shows how a student or child compares with their peers, while qualitative data provides details tailored to the student's unique environment, learning, and relationships. For example, a survey that is done at the beginning and the end of the year can indicate how a child's perceptions and coping strategies have changed as a result of any program that has been put into place during the year. This data can then be used to analyse if, and how, any programs have been effective in developing resilience skills, attitudes, and observable behaviours, within students in the learning environment. This type of data can become a very important aspect of the evaluation of a Makerspace as it highlights to key role this type of learning environment has in developing such a hard to measure trait.

Table 8.1 How resilience is reflected in the social capability learning continuum of the Australian curriculum. (© Australian curriculum and reporting authority [ACARA], 2020)

	Level 1	Level 2	Level 3	Level 4	Level 5	Level 6
	Foundation year	Year/Grade 2	Year/Grade 4	Year/Grade 6	Year/Grade 8	Year/Grade 10
Sub elements	Identify situations that feel safe or unsafe, approaching new situations with confidence/identify people and situations with which they feel a sense of familiarity or belonging	Undertake and persist with short tasks, within the limits of personal safety	Persist with tasks when faced with challenges and adapt their approach where first attempts are not successful	Devise strategies and formulate plans to assist in the completion of challenging tasks and the maintenance of personal safety	Assess, adapt and modify personal and safety strategies and plans, and revisit tasks with renewed confidence	Evaluate, rethink and refine approaches to tasks to take account of unexpected or difficult situations and safety considerations

Assessment processes can also highlight the importance of the symbiotic relationship between a child and their learning environment. Any form of measurement would take into consideration the following:

- the quality of relationships between children, their peers, and the significant adults they interact with;
- interventions that are enacted when children experience change and / or challenges (e.g. moving from middle to high school, coming to a new school, dealing with a death in the family, recovering from illness, etc.); and
- the capacity of the environment and educational program to develop a resilient skillset Beyond Blue (2017, pg. 34).

A Positive, Minus, and Improvement (PMI) tool or Strengths, Weaknesses, Opportunities, and Threats (SWOT) Analysis encourages students to reflect on both the positive and negative aspects in a task or situation and propose improvements. This type of reflective approach encourages students to consider their 'failures' as they can identify and improve on them.

In the Australian Curriculum, there are resilience-related outcomes that are built into the General Capabilities within the Personal and Social Capability Domain. Achievement levels are related to the Grades and Years and are discussed in more detail in Chapter 12. We have used the Case Studies below as the context in which these outcomes can be viewed in terms of observable student behaviours that you might see in your own classroom. Table 8.1 provides a brief overview of how resilience is described within various year levels, and Table 8.2 provides an example of what these outcomes might look like in the classroom.

Table 8.2 Mapping case study 8.1 resilient behaviours to the sub-outcomes of the social capability learning continuum. (© Australian curriculum and reporting authority [ACARA], 2020)

	Year/Grade 2	Year/Grade 4	Year/Grade 6
Sub elements	Undertake and persist with short tasks, within the limits of personal safety	Persist with tasks when faced with challenges and adapt their approach where first attempts are not successful	Devise strategies and formulate plans to assist in the completion of challenging tasks and the maintenance of personal safety
Behaviour	Behaviour was agitated and distressed when a task couldn't be completed immediately	James recognises that this task is difficult and fails his first attempt. He then pauses and makes another attempt to use the materials but is again unsuccessful	James recognises that this task is difficult after three attempts. He then seeks assistance from his friend and together they successfully complete the task

(continued)

Table 8.2 (continued)

	Year/Grade 2	Year/Grade 4	Year/Grade 6
Reflection questions	Did the task need to be broken down into pieces in order for James to complete it? What happened to James' behaviour when he got frustrated?	Did James complete the task with help/without help? What happened to James' behaviour when he got frustrated?	
Questions to ask student	What do you do when this doesn't work? How does this make you feel? Can you think what happens next? This must be frustrating for you but how can you keep going?	This looks tricky how are you going? What will you do next? Is there another way you can try? This must be frustrating for you but how can you keep going?	

8.8 Case Study

Most teachers are familiar with their students, and they can frequently predict or identify students that they know struggle in class and may evidence distress and confusion when they are not able to complete tasks (Fig. 8.2).

In 2017 and 2020, preservice teachers (PSTs) went to a school over a number of weeks to complete tasks with Grade 6 students in a one-on-one or two-to-one situation. The PSTs were asked to examine the learning and take non-identifying photos and audio of the students' experiences which we then discussed in the review back at the University.

Context 1. Fig. 8.3 (Makerspace Perth, 2017) represents a student struggling with their ability to persist at a task. The design brief was to develop and create an artefact (a Catapult) in Makerspace with the support of a preservice teacher (PST). In the foreground, you can see the pop sticks and elastic bands. In this task, students were required to create a catapult based on a sample provided on each table. They were given a fixed number of elastic bands and pop sticks and were required to examine the model to work out how to create their own identical item. This task was tricky as the tension on the frame of pop sticks needed to be maintained to keep the structure rigid. Any over-tightening of the elastic bands would cause the framework to move and a warping torsion, making it really challenging to add additional elastic bands.

When the photograph in Fig. 8.2 was taken, the Year 6 student had been working on his structure for 25–30 minutes and had only managed to connect three pop sticks out of the 12 that were provided. The PST in the blue 'STEMinist' t-shirt was very calm and did not interfere with the student. In fact, he just sat quietly next to the student and waited patiently until the student continued his construction. The PST was able to hold the structure to help the student, and the student did eventually manage to complete the task and make a catapult and he was very pleased that he had persisted.

Fig. 8.2 Kintsugi bowl

Fig. 8.3 Building resilience

In the class, the teacher had already indicated to the supervisor that this student often struggled with his persistence and would get distressed and cry in class if he was not able to complete the task. In this case, he used this moment to regroup and then he was able to demonstrate his resilience. His

PST encouraged him to keep trying and gave him time without comment to 'recover' and then he had continued. Had this been a teacher with a large class rather than a one-on-one experience, the student may have received help when he got distressed rather than being given the opportunity to work it out himself. By giving the student the time and space to work it out for himself, it helped build his resilience.

Context 2 Students in a 2020 Year 6 class were required to design and make a Grabber or use the Grabber design provided. Their task was to move a 1 litre bottle of water from one hoop on the floor to another hoop 3 meters away (Fig. 8.4). Students started by moving an empty bottle and then a bottle containing 100, 200, 300, 500 ml, and 1 L of water (Makerspace Perth. 2020) (Fig. 8.5).

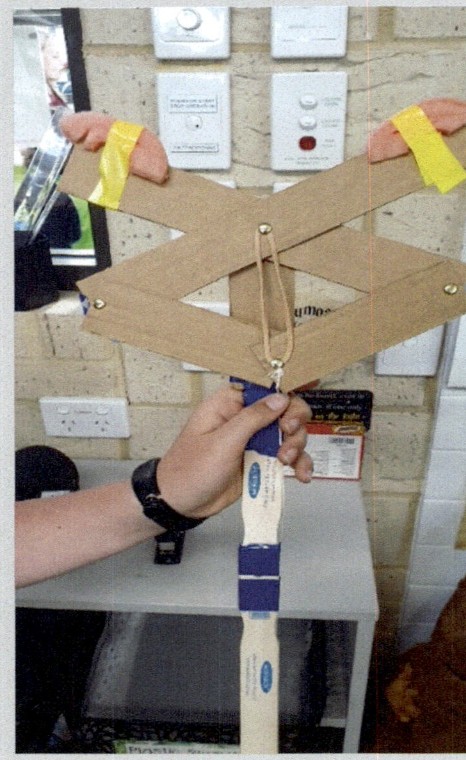

Fig. 8.4 Grabber design

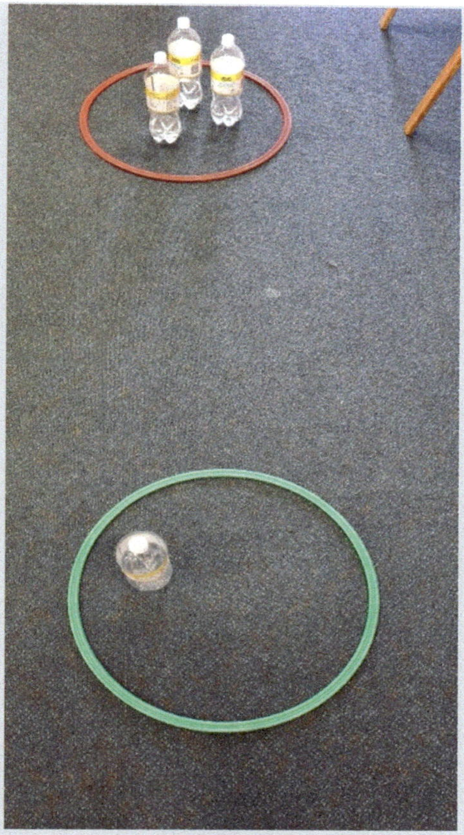

Fig. 8.5 Grabber testing area

One student complained to her teacher that she was finding it difficult to create a Grabber (Fig. 8.3). While most students persisted and refined their Grabber, she was sure from the beginning that she would not be successful. Even though she was typically a successful and high-achieving student, she made a single half-hearted attempt to move the bottle of water and then declared she was unable to complete the task.

The supervising PST reported surprise as in previous lessons the student had always been successful in attempting and completing tasks. While other students were engaged in testing their Grabber designs and watching others, this student was disengaged and was rebuked for talking several times (Makerspace Perth, 2020). This second example demonstrates a student struggling with resilience.

8.9 Summary

What is resilience?
The capacity to persevere through a challenge or recover quickly from a challenging event, situation, or experience.

Why is it important to develop resilience in children?
Students need to develop resilience in order to navigate the challenges they face both in and out of the classroom in order to take control of their own lives. Resilience gives individuals the confidence to face new challenges and uncertainty. According to Maslow's hierarchy of needs, resilience is an essential disposition needed to achieve self-actualisation.

How can a Makespace help develop resilience?
Makerspaces provide activities where students are not formally assessed. While this may challenge the value of this form of learning (because the activities are considered 'low stakes') students may in fact reach higher and further than they usually would reach because they are not hindered by fear of failure in the sense that it will affect their grades. Creating a supportive environment is extremely important. Teachers and students together or individually can decide what resilience is and then what criteria can be used to assess it.

What does resilience look like in a Makerspace?
Resilience can take many forms and will look different depending on the student and the activities being undertaken. Students' levels of resilience can also change depending on what is currently happening in their lives. In a Makerspace, resilience will be demonstrated by students who persevere through tasks until they are completed. Overcoming setbacks or failures will show that a student is exhibiting resilient behaviours, whether by themselves or with their peers.

How can resilience be assessed?
Resilience can be assessed using frameworks that examine students' behaviour through observations or through student self-reporting. Resilience has been identified and described within the sub-elements of the Personal and Social Capability General Capabilities from the Australian Curriculum, and this shows a developmental progress that can be achieved over the years as they progress through their schooling levels. The fact that these traits have been identified within formal curriculum documents implies they are assessable and reportable.

Appendix

Creating a Makerspace around Maslow's Hierarchy of Needs
In the table below, please consider what you can do in your Makerspace to create ensure Maslow's conditions are addressed.

Maslow's hierarchy condition	How would you create this in your Makerspace?
Survival	Example, Undertake a safety review of the physical space and check for defect and damage to the room including the furniture (desks and chairs), equipment (bookcases) and materials. Ensure all loose materials are stored correctly and there are no trip hazards or drop hazards anywhere
Security	
Belonging	
Belonging	
Respect	
Self-actualisation	

Part III
Country Context

Preface

Part III explores how existing global Makerspaces will influence their future iterations. We directly focus on the Australian and Indian Curricula and detail how relevant Makerspaces link with current educational imperatives. By assessing contemporary Makerspaces, we are able to analyse how and why certain Makerspaces are more successful than others and ultimately how to 'future-proof' Makerspaces for their continued success.

Chapter 9. "STEM, TVCs, and Makerspaces in the Indian Curriculum"

This chapter considers how science, technology, engineering, mathematics, and the twenty-first-century skills are considered within the Indian context in the current curriculum and what this may then look like in the future iteration of the curriculum in India being ideated by a team of experts from around the world.

Chapter 10. "STEM, TVCs, and Makerspaces in the Australian Curricula"

This chapter addresses how STEM is presented in the Australian Curriculum as well as Makerspace activities that link to the General Capabilities. By unpacking these capabilities, we are able to address their correlation with the Transversal Competencies explored in Section II and why their integration into contemporary pedagogy is vital in twenty-first-century learning.

Chapter 11. "Future-Proofing Makerspaces"

By looking at the transformation from Education 1.0 to Education 4.0, this chapter confirms Makerspaces as a necessity in sustaining education with the rapid technological innovation of the twenty-first century. The flexibility and openness of Makerspaces ensure their timeliness in the future of learning. This chapter also features several examples of virtual Makerspaces.

Chapter 9
STEM, TVCs, and Makerspaces in the Indian Curriculum

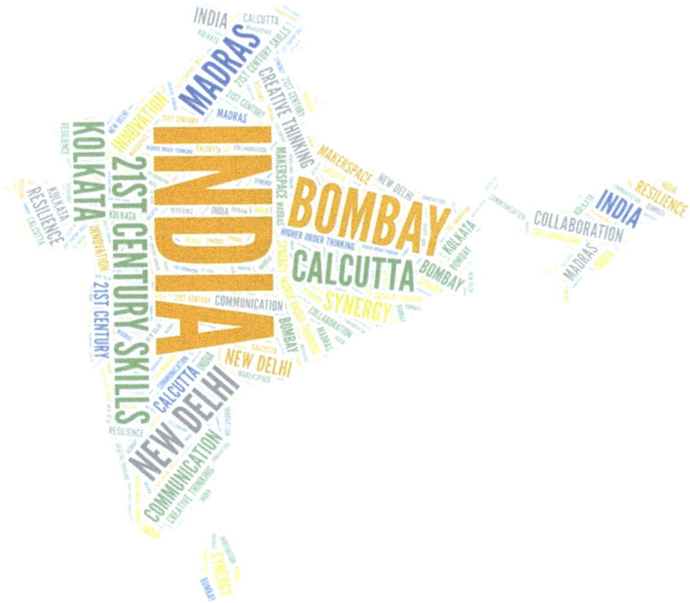

The Indian education system is one of the largest in the world with more than 1.5 million schools, 8.5 million teachers and 250 million children from varied socio-economic backgrounds (UNICEF, 2018).

Keywords National education policy · NCERT · Strands · Subject speciality

Definition of Indian Curriculum
The purpose of the education system is to develop good human beings capable of rational thought and action, possessing compassion and empathy, courage and resilience, scientific temper, and creative imagination, with sound ethical moorings and values. It aims at producing engaged, productive, and contributing citizens for building an equitable, inclusive, and plural society as envisaged by [the Indian] Constitution (NEP, 2020)

Focus Questions

- How is STEM taught in the Indian Curriculum?
- Which competencies are the Indian National Education Policy aiming to develop in students?
- Where are the Transversal Competencies identified in the Indian Curriculum?
- Where do Makerspaces fit in the Indian Curriculum?

R. B. Koul et al., *Teaching 21st Century Skills*,
https://doi.org/10.1007/978-981-16-4361-3_9

9.1 Introduction

India has a rapidly growing economy and has the second-largest population in the world of approximately 1.3 billion people. It has the highest number of youths, with the average age of 26 years, with 65% of the population below 35 years of age (UNDP, 2015). There are more than 1.5 million schools with over 260 million students enrolled, and, at a tertiary level, has approximately 864 universities, 40,026 colleges, and 11,669 institutes that cater for 3.57 million tertiary students (Koul, 2019). According to Sharma and Sharma (2015), education is organised by both State and Central Government. The Central Board of Secondary Education (CBSE) controls secondary exams in Classes X–XII and is supported by the National Council of Educational Research (NCERT), who is responsible for the Indian Curriculum, textbook production, and the dissemination of professional learning opportunities for teachers. It is also the key driver in policy and pedagogy development. Each State Board of Education also has their own Council of Educational Research and Training (SCERT).

Using the pipeline challenge as an example, we examine how STEM Makerspace activities reflect the curriculum outcomes articulated in key Indian curriculum documents: the *National Educational Policy* (NEP) 2020 and *Learning Outcomes at Elementary Stage*, 2017.

This chapter also examines the organisation of the primary curricula while considering how to embed STEM Makerspace activities into the learning program.

9.2 Transitioning Through the National Education Policy

India's National Education Policy has undergone major transitions since its first iteration in 1968. This policy proposed educational equality and free education until a child turns fourteen years of age. It was designed to be taught by specialised teachers in English, Hindi, and a variety of regional languages. Prime Minister Rajiv Gandhi launched the second NEP in 1986, which also sought to promote a child-centred approach and the creation of the rural University model. In 1992, there was a revision by the Government and addition of the Programme of Action (PoA), which is a common entrance examination across India for admission to professional and technical programs. Since then, there has been incredible transformation in the Indian school system, even before ushering in the new 2020 NEP aimed at systemic reform. This most recent document seeks to galvanise the Indian curriculum to nurture the development of knowledge acquisition, skills development, and attitudes that directly align with a commitment to human rights, and more specifically, sustainable living and global well-being. This directly aligns with Goal 4 of the 2030 Agenda for

Sustainable Development, which aims at *ensuring inclusive and equitable quality education and promoting lifelong learning opportunities for all Indian students by 2030* (© United Nations, 2020). This imperative requires the reorganisation and reprioritisation of the current education system in order to align learning with the goals of the 2030 Agenda for Sustainable Development. The NEP also seeks to reduce curriculum content to what will be termed the *core essentials* to enable teachers to develop learning spaces that promote the development of critical and creative thinking and display holistic learning. This is achieved by inquiry-focused discovery learning to encourage communication and collaboration. These articulated values and dispositions directly relate to the UN Transversal Competencies. Please refer to Chap. 4 which maps Transversal Competencies to a range of international curriculum documents.

The NEP 2020 is the first education policy of the twenty-first century that addresses the many growing developmental imperatives in India while emphasising the importance of individual and creative development. It outlines progressive education principles that aim to develop *foundational capacities* (literacy and numeracy), *cognitive capacities* (problem-solving and critical thinking), and *social, ethical, and emotional capacities.*

9.3 STEM in the Indian Curriculum

While the NEP 2020 espouses new pedagogical approaches, the Indian curriculum is still defined by *Learning Outcomes at the Elementary Stage, 2017*. This new document defines outcomes for 7 subjects: Hindi; English; Urdu; Mathematics; Environmental Studies; Science; and Social Sciences. For each of the subjects, pedagogical processes are suggested along with what learners should be able to do at the end of each class. Of the seven subjects taught in schools, Science and Mathematics are the only two subjects that fall within the realm of STEM (although the 'E' in STEM could be represented by the *Environmental Studies* rather than Engineering (Sheffield & Blackley, 2015)). Science outcomes are only stated for the upper primary (Classes VI to VIII), Mathematics outcomes are divided into two stages: Primary (Classes I–V) and Upper Primary (Classes VI to VIII). Prior to the introduction of the NEP 2020, Science was typically taught individually at lower secondary level as an integrated whole rather than a compartmentalised discipline (Ghosh, 2014). Discipline-oriented teaching and learning commences at XIth and XIIth grades, corresponding to students 16–18 years of age.

9.4 Science and Mathematics Learning Outcomes

The *Learning Outcomes at the Elementary Stage* describe science as an important human endeavour that can help students become effective problem solvers using scientific inquiry. This framework allows students to investigate authentic real-world problems such as sanitation, poverty, and sustainability. The Learning Outcomes of Mathematics brings up the difference between the curriculum and the learning outcomes by stating that:

> Curricular expectations define what a child should know… as well as the dispositions that should be acquired over a period of time. The learning outcomes [are] derived from the curricular expectations and… are generally treated as assessment standards or benchmarks for assessment. (Learning Outcomes at the Elementary Stage, 2017)

Both *National Education Policy 2020* and *Learning Outcomes at the Elementary Stage, 2017* are active steps towards a more transformational and timely education system. However, certain twenty-first-century programs are not well placed within the current Indian education system due to foundational shortcomings such as access to, and availability of, technology (as seen by the absence of *Engineering* and *Technology* in the curriculum) and the issue of knowledge-based learning against integrated education. Unlike the Australian curriculum that identifies Transversal Competencies in the General Capabilities (see Chap. 10) the Indian curriculum is yet to take the strategic step to ensure these competencies are transparent within curriculum documents and learning outcomes. The eventual integration of the NEP 2020 into the Indian curriculum, however, will usher in a highly innovative and twenty-first-century-focused education program.

The following example is a Makerspace activity that can be given to students to achieve certain prescribed learning outcomes identified in the NEP 2020. Part 1 focuses on content knowledge from Science, and Mathematics (due to the absence of Technology and Engineering learning outcomes) while Part 2 identifies twenty-first-century skills demonstrated when undertaking this activity.

9.5 Activity: Pipeline Challenge—Part 1

Part 1 of this example focuses on the Science and Mathematics content knowledge addressed in the Pipeline Challenge activity and how it can be assessed in different class levels. As mentioned previously, Technology and Engineering, they are unable to be appropriately mapped to curriculum outcomes. This activity is low-cost and can be adapted by teachers in resource-limited schools to teach elementary-level students in an engaging way.

Setting the Scene:

- Nearly 76 million people in India do not have access to safe drinking water.

Table 9.1 Ideation questions

Questions	Answers
How much water is in a cubic metre?	1000 L = 1 cubic metre
How long does it take for water to flow?	It depends on the distance to travel, pressure with which it is pumped and number of users
What makes the water move from dam to place of supply?	Pumps force the water to move, a bit like when you spit water out of a tube by blowing at one end. Water likes to flow downhill because of gravity, the same way you drop something and it falls to the ground. Dams are generally located high in the hills, but sometimes water has to be supplied to higher grounds. Apart from moving the water along, the pumps have to push the water up to a higher elevation
Why have pipes and not a channel?	A channel is open to the air and cannot be pressurised. A pipe can be pressurised and then water can be pumped uphill. Water can evaporate from a channel. A channel could be polluted whereas a pipe is sealed and protected
What do you think pipes are made of?	Originally pipes were made of steel plates. In the 1930s, pipes were lined with concrete to overcome some corrosion problems. What are todays pipelines made of? Students to suggest or discover
What other things can be transported using pipelines?	Water, oil, gas, sewerage, chemicals in a chemical plant, air as in air-conditioning, concrete, powders, blood (dialysis), steam, etc

- 54% of Indian rural women have to travel between 200 m and five km daily to get drinking water. As a result, they walk 20 min in a day on an average and spend another 15 min at the water source.
- Collectively these women cover 64,000 times the distance between the earth and the moon (wearewater.org) (Table 9.1).

9.6 Activity: Pipeline Challenge

Context
Students will design a system prototype for transporting water over long distances with a predetermined set of tools and equipment. Water is represented by a table tennis ball in this activity.

Materials
- One table tennis ball (1 per group for testing but also to be used for prototyping as pipeline is designed and built);
- Six A3 sheets or newspaper suitable for making pipes and special joins;
- Packing tape or sticky tape and scissors;
- Two-metre tape measure;
- One protractor to measure angles (students can be encouraged to make their own protractor. A corner of paper is 90 degrees and a corner evenly folded into 3 equal angles is 30 degrees).

Instructions
Student-designed pipelines must be able to transport a single table tennis ball the length of the pipe without external assistance.

1. Pipe length is two metres—this can be modified to increase or decrease the complexity of the activity.
2. The pipeline must have four angles to represent the pipeline's requirement of moving around environmentally protected areas. One angle must be 90 degrees while the other three must be at least 30 degrees.
3. Show students the path they need to traverse in the classroom.
4. The starting point will be 50 cm above the ground (chair height) and the finish point will be floor level.
5. The only materials to be used are cardboard, paper, and tape, or other alternatives chosen by the teacher.
6. Allow ~ 20 min for students to develop their pipeline design on paper. This time could also be used for prototyping if desired. The final plan should include:

a diameter
b The length of each section between angles?
c how sections will be joined?
d how the angle connections will be designed and built?

7. Pipeline construction (~one hour) (Figs. 9.1 and 9.2).

Fig. 9.1 Prototype of pipeline challenge construction part 1

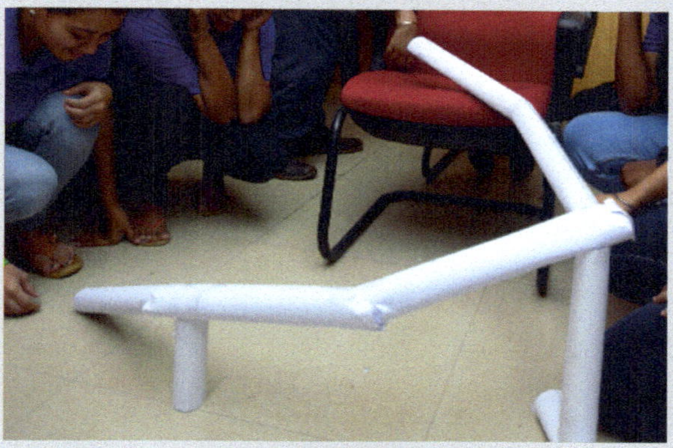

Fig. 9.2 Prototype of pipeline challenge construction part 2

Evaluation and Teacher's role

Students will need to continually re-evaluate their design during the construction process, particularly the angle connections. Teachers should encourage students to test their prototypes to identify design and construction issues early. Each pipeline is evaluated and scored against the project marking criteria which can be changed by the teacher depending on modifications to the activity.

Activity Marking Rubric

- What was the hardest part of designing and building your pipeline?
- What parts of your pipeline did you have to re-engineer?
- Which pipeline worked best and why?

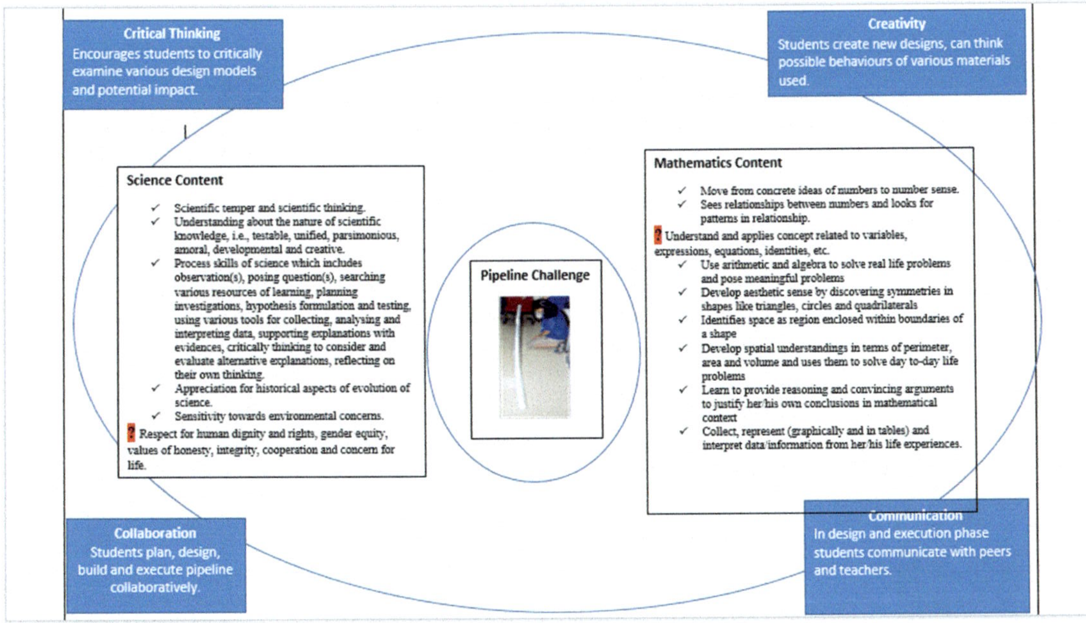

Fig. 9.3 Pipeline activity demonstrating transversal competencies

9.7 Pipeline Challenge: Links to Science Curriculum

Of the eight themes in the science curriculum, the pipeline challenge involves the following six: Food, Materials, People and Ideas, How Things Work, Natural Phenomena, and Natural Resources (Tables 9.2, 9.3).

Table 9.2 Marking rubric

	Build			Commissioning		
Team name	Length 2 M (2 point/M)	90° angle (3 points)	3 Angles > 30° (2 Points for each angle)	Table tennis ball (3 points /m)	Ball passes at angles (2 points/angle)	Total points
Team A						
Team B						
Team C						

Table 9.3 Science curriculum outcomes achieved in the pipeline challenge

Curricular expectation	Content covered
Scientific temper and scientific thinking	Design, construction, and execution of pipeline is fostering scientific temper at each stage
Understanding different types of scientific knowledge (factual, theoretical, etc.)	Students are introduced to the issue of transporting fluids, e.g. water, oils, etc., in safe way
Apply science skills such as observation, research and recording, investigation, prototyping and testing, and collecting and analysing data and figures	Teachers are provided with suggestions on how to encourage students to observe the phenomena, ask questions, design, and redesign, test the output
Appreciation for historical aspects of the evolution of science	Development of pipes and changes in materials used
Sensitivity towards environmental concerns	Access to safe drinking water is a major environmental issue in India Safe transportation of other hazardous materials through pipelines is also covered
Respect for humanity and understanding of equal rights in gender, age, sexuality, and disability, along with integrity, cooperation, and dignity	The last set of Curricular Expectations is closely linked and discussed in the section on the attainment of twenty-first-century skills

9.8 Pipeline Challenge: Links to Mathematics Curriculum

The following learning outcomes in mathematics are divided into the Elementary Stage and Upper Primary Level (Table 9.4).

9.9 The NEP 2020 and Makerspaces

As the current curriculum reflects the 1992 NEP, there still remains a fixed pedagogy where students are passive participants in their learning.

In this static classroom climate, Makerspaces and a Makerspace approach are incongruent with current practices. As discussed in Chap. 1, Makerspaces promote a strong play-based hands-on exploratory approach to learning, and therefore, Makerspaces would be sidelined activities in the Indian curriculum due to the heavy content-focus that does not explicitly emphasise more organic learning opportunities. Makerspace activities can facilitate an environment where students feel welcomed and cared for, where a safe and stimulating learning environment exists, but the curriculum must reflect the opportunity to do so, followed by the implementation and upkeep of Makerspace. The new NEP 2020 document has articulated, for the first time, the importance of twenty-first-century skills in the life of Indian students. It has an overarching emphasis on critical thinking, creativity, collaboration, communication, and a number of other important values and dispositions—one of the eight key principles of the NEP (2020) is *critical thinking and creativity to boost logical decision-making and innovation.* The NEP 2020 has reduced the number of core curriculum concepts to enable teachers to spend more time teaching skills rather than only content. This, however, comes with the issue of teachers having to adjust their personal pedagogical approach.

The deciding factor in curriculum integration of the most recent NEP document is how it is translated from the context of overarching ideas into classroom practice. Teachers will require a curriculum that specifies how they may assess the skill, similar to Australia, the UK, the USA, and parts of Europe. This is seen in Chap. 10: Table 10.5, where elements and sub-elements of the General Capabilities (observable behaviours) are identified on a continuum through all grades in primary and secondary classes. If the Indian curriculum wants to teach similar values and dispositions, then a new continuum is needed to enable Indian teachers to identify, map, and then assess them. Another issue will be helping Indian teachers to adopt a new approach to their teaching which encourages students to be more hands-on and engaged in their learning. Teachers will be required to set up learning intentions that encourage students to demonstrate their critical thinking and also their creativity. Extensive Professional Development will be needed to support educators enabling them to structure the new learning.

Table 9.4 Mathematics curriculum outcomes achieved in the pipeline challenge

Curricular expectation	Content covered
Elementary stage	
Cultivate a connection between mathematical thinking and day-to-day life contexts	Students can calculate the amount of water being used by them or their families on daily basis
Understand shapes and articulate their observable properties as similarities and differences among them	Activity gives an understanding of circular items
Develop own methods of performing operations on numbers in daily life (addition, subtraction, multiplication, and division)	All four basic mathematical operations can be taught to students in different stages of the project, i.e. planning, building, and execution
Develop language and symbolic notations with standard algorithms of performing number operations	Teacher can incorporate the use of mathematical operations language (addition, subtraction, multiplication, and division) and associated symbols
Estimate outcome of operations on two or more numbers and use it in daily life activities	Calculations can be based on the level and grade of the students. Using two number operations can be easily incorporated
Learn to represent the part of a whole as a fraction and order simple fractions	A comparison between liquid carried through a wider and narrower pipeline can be calculated to introduce the concept of fraction
Collate and interpret data from daily contexts	This activity is of direct relevance to everyday life, thus an authentic learning experience
Identify and extend simple patterns in shapes and numbers	Simple patterns like rods, cones, circles are introduced to students
Upper primary level	
Move from concrete ideas of numbers to number sense	The level of difficulty in the activity can be increased with the level of students undertaking pipeline activity and difficulty level for ideas of numbers
Sees relationships between numbers and looks for patterns in relationship	Students are introduced to concept of area in relation to shapes and patterns
Understand and applies concept related to variables, expressions, equations, identities, etc.	Calculating output through pipeline for given outcomes is explained below: Variables—transportation of liquids, e.g. milk, water, oil Expressions—mathematical expressions like angels, lengths, and diameter Equations—students learn to calculate quantity of fluids passing through a pipeline

(continued)

Table 9.4 (continued)

Curricular expectation	Content covered
Use arithmetic and algebra to solve real-life problems and pose meaningful problems	Concept of line segment, lines, parallel lines, angles—acute, right, obtuse, complementary, supplementary, adjacent, and vertical, triangles, congruent and similar figures, circle, radius, diameter and chord, arcs, semicircle and segments of circle, built from papers, worksheets are integrated
Learn to provide reasoning and convincing arguments to justify her/his own conclusions in mathematical context	Active discussion by setting the scene and asking prompting questions is encouraged
Collect, represent (graphically and in tables) and interpret data/information from her/his life experiences	Activity marking rubric provides representing data in tables and can further be transformed into graphs and charts

9.10 Makerspace is not Jugaad

Makerspaces can be used as a mechanism to support Indian teachers as they create an environment where students feel safe and encourage them to take risks and not fear failure. Makerspace *space* encourages students to play and to be hands-on and explorative in their practice. There is concern that the Indian Makerspace is in danger of being confused with the traditional notion of Jugaad of which India has a long tradition (Verma, 2019). The term '**Jugaad**' is a colloquial Hindi word, which approximately translates as 'quick fix', 'workaround', or 'hack'. Jugaad thinking is a much more haphazard solution that solves the problem often in a chaotic or non-systematic way, especially when the problem-solver lacks the skills and understanding. Makerspaces are the antithesis of Jugaad: they strive for a scientific systematic design thinking approach. Makerspaces look at all the solutions, and the one chosen is chosen because it is the best solution (and not just the most efficient) and can be rationalised in STEM content knowledge within the engineering, technology, science, and mathematics.

9.11 Summary

How is STEM taught in the Indian Curriculum?

At present, STEM subjects (science and mathematics) are taught in a siloed manner. The Indian curriculum focuses on content, thus making deviations from assessment-driven pathways difficult for teachers. Examinations and assessments have typically been key drivers of the curriculum and a measure of student success. It is hoped that the NEP 2020 will reduce the strict emphasis placed on content knowledge and promote a more holistic and integrated Indian curriculum. This could potentially allow STEM Makerspaces to occupy a more mainstream position in classrooms and integrate engineering and technology once the NEP 2020 has been adopted into contemporary Indian curriculum.

Which competencies are the Indian National Education Policy aiming to develop in students?
In line with the focus of Sect. 9.2 of this book, the NEP 2020 document articulates the importance of four twenty-first-century learning capabilities in students: critical thinking, creativity, collaboration, and communication. These are specifically espoused in NEP (2020) document and could be supported through a Makerspace program.

Where are the Transversal Competencies situated in the Indian Curriculum?
The Transversal Competencies, defined more prominently as *values* and *dispositions* throughout Indian education documents, are found throughout the National Education Policy 2020. It is clear that the curriculum and policy developers are aware of the value and importance of these competencies which align with the policy and rhetoric in Europe, USA, Australia, and the UK and from reviews around future needs by employers. However, at this stage of integration with the NEP, they do not explicitly appear to have a prominent place within curriculum documents or described in behavioural outcomes.

Where do Makerspaces fit into the Indian Curriculum?
The STEM pipeline activity demonstrates that STEM components can be facilitated in early, middle, and late elementary levels, and a skilled educator can create connections and encourage students to attempt to explore real-world problems such as water flow and the need for water. Unlike Jugaad, Makerspaces promote the systematic development of design principles and fluidity that includes collecting data, testing, and communicating. As explored in this chapter, there is proof that within the new Indian National Education Policy (NEP) 2020, Makerspaces can fit into regular classrooms and be part of the suite of pedagogical tools and approaches Indian teachers have to enable them to teach content knowledge and the stated values and dispositions. This is to support students as lifelong learners capable of critical and creative dispositions and temperaments that are supportive of solving the world's problems.

Chapter 10
STEM, TVCs, and Makerspaces in the Australian Curricula

More important than the curriculum is the question of the methods of teaching and the spirit in which the teaching is given Bertrand Russell (Sejnost, 2009).

Keywords General capabilities · Content descriptor · Strands · Subject speciality · STEM · Transversal competencies

Definition of Australian Curriculum
The Australian Curriculum sets consistent national standards to improve learning outcomes for all young Australians. It defines, through content descriptions and achievement standards, what students should be taught and achieve as they progress through school. It is the base for future learning, growth, and active participation in the Australian community (© Australian Curriculum, Assessment and Reporting Authority [ACARA], 2020).

Focus Questions

- How is STEM taught in the Australian Curriculum?
- What are the General Capabilities in the Australian Curriculum?
- Where are the Transversal Competencies situated in the Australian Curriculum?
- Where are Makerspaces placed in the Australian Curriculum?

R. B. Koul et al., *Teaching 21st Century Skills*,
https://doi.org/10.1007/978-981-16-4361-3_10

10.1 The Australian Curriculum

The Australian Curriculum, Assessment and Reporting Authority [ACARA] is the body responsible for the development, implementation, and ongoing review of the Australian Curriculum. Since 2014, all Australian states and territories have implemented a Foundation to Year 10 Curriculum, either directly aligned with the National Curriculum or a modified, state-specific version. According to the ACARA (2020), over 10,000 schools nationwide use the National Curriculum. It was developed after extensive consultation of nearly 10,000 (9787) participants and review of international and national curricula and considers the learning needs of the twenty-first century and educational goals for young Australians. Teacher consultants were co-opted from all states and territories to develop the draft curriculum over eighteen months with further review of 14,816 submissions. The draft curriculum was then trialled by 2513 teachers in classrooms and finally approved by the ACARA board and all Education Ministers around Australia. In 2016, it was reported that 3,798,266 students had accessed this new curriculum.

The Australian Curriculum is made up of eight learning areas: English; Mathematics; Science; Humanities; Social Sciences; The Arts; Technologies; Health and Physical Education; and Languages. They are taught in strands and sub-strands from Foundation to Year 10, with students having the option to participate in a further two years of a senior secondary curriculum. The focus here will be on the F-Year 6, considered to be the Primary School, which is further divided into early childhood (Foundation to Grade 2) and primary education (Grades 3–6).

10.2 STEM in the Australian Curriculum

STEM in the Australian Curriculum is identified as a combination of Science, Technology, and Mathematics. Engineering is not taught as a subject, but aspects of the discipline can be seen in the Science Inquiry strand and in the Technology Learning area. Below we discuss the Science Learning Area in depth and briefly summarise Technology and Mathematics.

Science

In Science, there are three strands:

- Science Understanding;
- Science as a Human Endeavour; and
- Science Inquiry Skills.

In the Science Understanding strand, there are four sub-strands:

- Biological Science;
- Chemical Science;

- Earth and Space Science; and
- Physical Science.

For each sub-strand in each year, there are content descriptors that indicate which concept should be taught. While the teacher determines how these concepts will be introduced into the classroom, they must adhere to the achievement standard stipulated. All four strands must be taught within a one-year timeframe, which gives the teacher one quarter of the year (one school term in Australia) to teach all aspects of a sub-strand. As students' progress through the different years, the concepts presented in each strand are developmental and designed to build on key understandings taught in the previous year(s). Table 10.1 shows the progression of the Science Understanding sub-strands from Years 3 to 6 and the increasing complexity of each topic.

Within the Science as a Human Endeavour strand, there are two sub-strands; the **Nature and development of science** and **Use and influence**. The content descriptors

Table 10.1 Science understanding content descriptors (mastered over one year) (© Australian Curriculum and Reporting Authority [ACARA], 2020)

Earth sciences	Biological sciences	Chemical sciences	Physical sciences
Year 3 *Earth's rotation on its axis* causes regular changes, including night and day (ACSSU048) **Year 4** *Earth's surface* changes over time as a result of natural processes and human activity (ACSSU075) **Year 5** The Earth is part of a *system of planets* orbiting around a star (the sun) (ACSSU078) **Year 6** Sudden geological changes or *extreme weather* conditions can affect Earth's surface (ACSSU096)	**Year 3** *Living things* can be grouped on the basis of observable features and can be distinguished from *non-living things* (ACSSU044) **Year 4** Living things have *life cycles* (ACSSU072) Living things depend on each other and the environment to survive (ACSSU073) **Year 5** Living things have structural features and *adaptations* that help them to survive in their environment (ACSSU043) **Year 6** The growth and survival of living things are affected by the *physical conditions of their environment* (ACSSU094)	**Year 3** *A change of state* between solid and liquid can be caused by adding or removing heat (ACSSU046) **Year 4** *Natural and processed materials* have a range of physical properties that can influence their use (ACSSU074) **Year 5** *Solids, liquids, and gases* have different observable properties and behave in different ways (ACSSU077) **Year 6** *Changes to materials* can be reversible or irreversible (ACSSU095)	**Year 3** *Heat* can be produced in many ways and can move from one object to another (ACSSU049) **Year 4** *Forces* can be exerted by one object on another through direct contact or from a distance (ACSSU076) **Year 5** *Light* from a source forms shadows and can be absorbed, reflected, and refracted (ACSSU080) **Year 6** *Electrical energy* can be transferred and transformed in electrical circuits and can be generated from a range of sources (ACSSU097)

Table 10.2 Science as a human endeavour (mastered over two years). (© Australian curriculum and reporting authority [ACARA], 2020)

Nature and development of science	Use and influence
Year 3,4 Science involves making predictions & describing patterns and relationships (ACSHE050) & (ACSHE061) Year 5,6 Science involves testing predictions by gathering data and using evidence to develop explanations of events and phenomena and reflects historical and cultural contributions (ACSHE081) & (ACSHE098)	Year 3,4 Science knowledge helps people to understand the effect of their actions (ACSHE051) & (ACSHE062) Year 5,6 Scientific knowledge is used to solve problems and inform personal and community decisions (ACSHE083) & (ACSHE100)

are taught and mastered over two years. The Year 3,4 and the Year 5,6 content descriptors are listed in Table 10.2 below. These broad statements can be adapted into a wide range of contexts.

Within the Science Inquiry Skills, there are six sub-strands which include **Questioning and Predicting, Planning and Conducting, Processing and Analysing Data and Information,** and **Evaluating and Communicating.** The content descriptors are also taught and mastered over two years. The Year 3,4 and the Year 5,6 content descriptors are listed in Table 10.3. These statements encourage teachers to think about the inquiry process within the context of science content. For example, Year 5 students would undergo a scientific investigation to find out about the solar system (ACSSU078).

Mathematics
In Mathematics, there are three strands:

- Number and Algebra;
- Measurement and Geometry; and
- Statistics and Probability.

Each strand has content descriptors that need to be mastered in one year and like the Science content descriptors progress along a developmental continuum. They also have proficiency elements that are used to assess a student's **fluency, reasoning, problem-solving,** and **understanding** in relation to each strand.

Technology
Technology describes knowledge and skills to design, create, and ideate and also to make informed decisions about the use of technologies in the environment and economy for a sustainable future. Technologies incorporate two distinct but related subjects:

Table 10.3 Science inquiry skills (mastered over two years). (© Australian curriculum and reporting authority [ACARA], 2020)

	Year 3,4	Year 5,6
Questioning and predicting	With guidance, identify questions in familiar contexts that can be investigated scientifically and make predictions based on prior knowledge (ACSIS53) & (ACSIS64)	With guidance, pose clarifying questions make predictions about scientific investigations (ACSIS231) & (ACSIS232)
Planning and conducting	With guidance, plan and conduct scientific investigations to find answers to questions, considering the safe use of appropriate materials and equipment (ACSIS054) & (ACSIS65) Consider the elements of fair tests and use formal measurements and digital technologies as appropriate, to make and record observations accurately (ACSIS055) (ACSIS066)	Identify, plan, and apply the elements of scientific investigations to answer questions and solve problems using equipment and materials safely and identifying potential risks (ACSIS086) & (ACSIS103) Decide variables to be changed and measured in fair tests & observe, measure and record data with accuracy using digital technologies as appropriate (ACSIS087) & (ACSIS104)
Processing and analysing data and information	Use a range of methods including tables & simple column graphs to represent data, to identify patterns and trends (ACSIS057) & (ACSIS068) Compare results with predictions, suggesting possible reasons for findings (ACSIS215) & (ACSIS216)	Construct and use a range of representations, including tables & graphs, to represent & describe observations, patterns, or relationships in data using digital technologies as appropriate (ACSIS090) & (ACSIS107) Compare data with predictions and use as evidence in developing explanations (ACSIS218) & (ACSIS221)
Evaluating	Reflect on investigations, including whether a test was fair or not (ACSIS058) & (ACSIS69)	Reflect on and suggest improvements to scientific investigations (ACSIS091) & (ACSIS108)
Communicating	Represent and communicate observations, ideas, and findings using formal and informal representations (ACSIS060) & (ACSIS71)	Communicate ideas, explanations, and processes using scientific representations in a variety of ways, including multimodal texts (ACSIS093) & (ACSIS110)

- **Design and Technologies**, in which students use design thinking and technologies to generate and produce designed solutions for authentic needs and opportunities, and
- **Digital Technologies**, in which students use computational thinking and information systems to define, design, and implement digital solutions (Australian Curriculum, Assessment and Reporting Authority [ACARA], 2020).

Design and Technologies have two sub-strands:

- Knowledge and Understanding (4 content descriptors); and,
- Processes and Production Skills (4 content descriptors).

Digital Technologies also have two sub-strands:

- Knowledge and Understanding (2 content descriptors); and
- Processes and Production Skills (7 content descriptors).

These content descriptors should be mastered over 2 years, in Year 3,4 and then Year 5,6.

Engineering
There is no Engineering subject in the Australian Curriculum in Years 1–10, but aspects of the design process can be seen in the *Science Inquiry Skills* strand and in the Design and Technologies *Process and Production Skills* sub-strand.

10.3 Makerspace Activities that Link to the Australian Curriculum

Designing a Wigglebot is an example of a common Makerspace activity which can be directly linked to a range of outcomes from the STEM subjects of Science, Technology, and Mathematics. We have identified direct links to content descriptors for Years 5 and 6. Part 1 focuses on content knowledge from Science, Mathematics, and Technology, emphasising the relevant sub-strand, content descriptor, while Part 2 identifies Transversal Competencies (referred to as General Capabilities in the Australian Curriculum) demonstrated when undertaking this activity.

10.4 Wigglebot Example Part 1

Students use their understanding of circuits to connect a motor to a Wigglebot to make it move. This activity addresses Year 5,6 content descriptors from Science, Mathematics, and Technologies.

Links to Science

Science Understanding *Electrical energy can be transferred and transformed in electrical circuits and can be generated from a range of sources* (ACSSU097):

- recognising the need for a complete circuit to allow the flow of electricity; and
- exploring the features of electrical devices such as switches and light globes (Fig. 10.1).

Science as a Human Endeavour *Scientific knowledge is used to solve problems and inform personal and community decisions* (ACSHE083):

- considering how electricity and electrical appliances have changed the way some people live; and
- discussing the use of electricity.

Science Inquiry Skills *Identify, plan, and apply the elements of scientific investigations to answer questions and solve problems using equipment and materials safely and identifying potential risks by* (ACSIS086):

- following a procedure to design an experimental or field investigation; and
- discussing methods chosen with other students, and **refining methods** accordingly (Fig. 10.2).

Adjusting the peg changes the balance line and rotating the peg moves the balance and generates the wiggle. This can be investigated and refined, and students can

Fig. 10.1 Simple electrical circuit

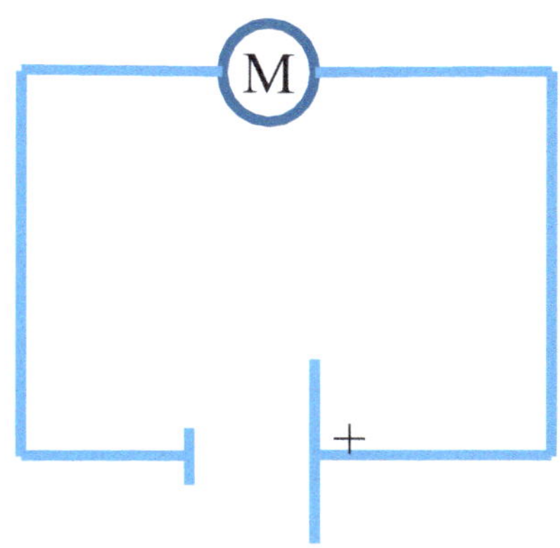

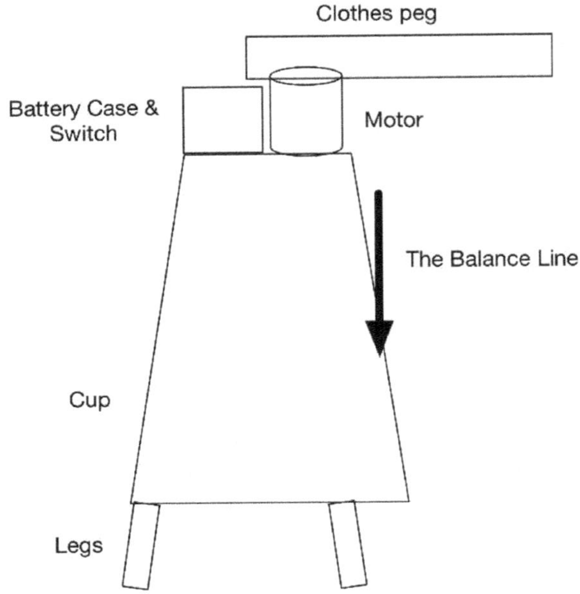

Fig. 10.2 Diagram of a Wigglebot

experiment to move the Wigglebot a little, a small wobble, to a big wobble that makes the bot move. Patterns are created on the paper with the pen lids removed.

Links to Mathematics

Both of the content descriptors below can be assessed through the examination of the pen legs of the Wigglebot. Students have to situate three pen legs around the rim of the cup in order to balance the Wigglebot and ensure that it can move. Questions either written or in an interview can ask how students position the pens and what they think the best position(s) is.

Measurement and Geometry *Estimate, measure, and compare angles using degrees. Construct angles using a protractor* (ACMMG112).

Investigate, with and without digital technologies, angles on a straight line, angles at a point, and vertically opposite angles. Use results to find unknown angles (ACMMG141).

Links to Digital Technology

If you wish to assess technology outcomes, the project could be extended to address the following content descriptor: *Investigate characteristics and properties of a range of materials, systems components, tools and equipment, and evaluate the impact of their use* (ACTDEK023). This would direct the focus of the activity to the materials used to construct the Wigglebot and allow students to determine if using other materials would have the same or a better outcome (Australian Curriculum, Assessment and Reporting Authority [ACARA], 2020).

10.5 The Place of Transversal Competencies in the Australian Curriculum

> The Australian Curriculum identifies seven capabilities that play a significant role in equipping young Australians to live and work successfully in the twenty-first century" (Australian Curriculum, Assessment and Reporting Authority [ACARA], 2020).

The Transversal Competencies identified in UNESCO's 2015 report are represented in the Australian Curriculum as General Capabilities and can be seen more explicitly across multiple subjects' content descriptors and are embedded in the content of different learning areas. Gonski's 2018 report also uses the term 'general capabilities' and refers to a 'clear list that has been nationally agreed and established as part of the Australian Curriculum' (Gonski, 2018). These General Capabilities can be placed into five categories of skill focus: skills related to **self** (self-discipline, independent learning, flexibility, adaptability, self-awareness, perseverance, self-motivation, compassion, integrity, initiative, risk-taking, self-respect, sense of belonging); skills related to **working with others** (presentation skills, communication skills, leadership, organisational skills, teamwork, collaboration, conflict resolution, sociability, collegiality, empathy, compassion); skills related to **global perspective** and **understanding** (awareness, tolerance, openness, respect for diversity, intercultural understanding, civic / political participation, respect for the environment, national identity); skills related to **critical and creative thinking** (creativity, entrepreneurship, resourcefulness, application skills, reflective thinking, decision-making); and, finally, skills related to **ICT and media literacy** (accessing information, locating information, communicating ideas, participating in democratic processes, analysing information and media, evaluating information and media content). When compared to UNESCOs Transversal Competencies the skills clusters show a close alignment between the two frameworks.

The Australian Curriculum General Capabilities, unlike the Transversal Competencies, are already developed into detailed sub-categories. It is posited, therefore, that 'the Australian Curriculum implicitly and explicitly includes transversal competencies in every educational activity. To this end, the Australian Curriculum provides detailed information on each capability and how it can be adopted across each subject' (UNESCO, 2015). Table 10.4 details the commonalities between the UNESCO Transversal Competencies and the General Capabilities defined in the Australian Curriculum (ACGC).

Table 10.4 Language used to describe the UNESCO transversal competencies and the general capabilities. (© Australian curriculum and reporting authority [ACARA], 2020)

UNESCO TVC	Key skills and competencies	ACGC
Critical innovative thinking	Creativity, entrepreneurship, resourcefulness, application skills, reflective thinking, decision-making	Critical and creative thinking
Interpersonal skills	Presentation skills, communication skills, leadership, organisational skills, teamwork, collaboration, initiative, sociability, collegiality empathy, compassion	Personal and social capability
Intrapersonal skills	Self-discipline, independent learning, flexibility, adaptability, self-awareness, perseverance, self-motivation, compassion, integrity, risk-taking, self-respect, sense of belonging	Personal and social capability
Global citizenship	Awareness, tolerance, openness, respect for diversity, intercultural understanding conflict resolution, civic / political participation, respect for the environment, national identity	Intercultural understanding
Media and information literacy	Accessing information, locating information, communicating ideas, participating in democratic processes, analysing information and media, evaluating information and media content	ICT capability critical and creative thinking

10.6 Unpacking the General Capabilities

In the Australian Curriculum, there are seven General Capabilities:

- Literacy;
- Numeracy;
- Information and Communication;
- Critical and Creative thinking;
- Personal and Social capability; and
- Ethical Understanding.

Table 10.5 Critical and creative thinking deconstructed into elements and sub-elements. (© Australian curriculum and reporting authority [ACARA], 2020)

General capability	Element	Sub-elements
Critical and creative thinking	Inquiring—identifying, exploring, and organising information and ideas	• Pose question • Identify and clarify • Organise and process information
	Generate ideas, possibilities, and actions	• Imagine possibilities and connect ideas • Consider alternatives • Seek solutions and put ideas into action
	Reflecting on thinking and processes element	• Think about thinking (metacognition) • Reflect on processes • Transfer knowledge into new contexts
	Analysing, synthesising, and evaluating reasoning and procedures element	• Apply logic and reasoning • Draw conclusions and design a course of action • Evaluate procedures and outcomes

Each of these General Capabilities describes observable behaviours which allow teachers to make judgements of a student's ability to, for example, 'reflect on processes' and 'apply logic and reasoning'. Table 10.5 provides an example of how a General Capability (e.g. critical and creative thinking) can be deconstructed into elements and sub-elements (observable behaviours that can be assessed).

Table 10.6 expands upon the sub-elements identified in Table 10.5 and shows their progression from Foundation to Year 10 (through six levels).

When assessing Transversal Competencies, it is noted that they can be found across a number of General Capabilities in the Australian Curriculum. Higher-order thinking has an obvious link to the Critical and Creative Thinking GC; however, other TVCs such as resilience, which is described in detail in Chap. 8, are less easily identified. Resilience is most directly linked to the sub-element **Become confident, resilient, and adaptable** under the *self-management* domain in the *Personal and Social Capability* General Capability. Table 10.7 shows the observable student behaviours linked to resilience through Levels 2, 3, and 4.

Table 10.8 expands upon the *Personal and Social Capability* GC by providing an example of a student's development of resilience, followed by Table 10.9, which lists examples of student and teacher reflective questions.

Table 10.6 Levels of general capability critical and creative thinking elements and sub-elements (Australian Curriculum, assessment, and reporting authority [ACARA], 2017)

Sub element	Pose questions	Identify clarify information	Organise and process information
	Inquiring—identifying, exploring, and organising information and ideas element		
Level 1 Foundation Year	Pose factual and exploratory questions based on personal interests and experiences	Identify and describe familiar information and ideas during a discussion or investigation	Gather similar information or depictions from given sources
Level 2 Year 2	Pose questions to identify and clarify issues, and compare information in their world	Identify and explore information and ideas from source materials	Organise information based on similar or relevant ideas from several sources
Level 3 Year 4	Pose questions to expand their knowledge about the world	Identify main ideas and select and clarify information from a range of sources	Collect, compare, and categorise facts and opinions found in a widening range of sources
Level 4 Year 6	Pose questions to clarify and interpret information and probe for causes and consequences	Identify and clarify relevant information and prioritise ideas	Analyse, condense, and combine relevant information from multiple sources
Level 5 Year 8	Pose questions to probe assumptions and investigate complex issues	Clarify information and ideas from texts or images when exploring challenging issues	Critically analyse information and evidence according to criteria such as validity and relevance
Level 6 Year 10	Pose questions to critically analyse complex issues and abstract ideas	Clarify complex information and ideas drawn from a range of sources	Critically analyse independently sourced information to determine bias and reliability

Table 10.7 Different ways resilience is described within the *Personal and Social Capability*. (© Australian curriculum and reporting authority [ACARA], 2020)

	Level 2	Level 3	Level 4
	Year 2	Year 4	Year 6
Sub element—become confident, resilient, and adaptable	Undertake and *persist* with short tasks, within the limits of personal safety	*Persist with tasks* when *faced with challenges* and *adapt their approach* where first attempts are not successful	*Devise strategies* and formulate plans to assist in the *completion of challenging tasks* and the maintenance of personal safety

Table 10.8 Student example of resilient behaviours. (© Australian curriculum and reporting authority [ACARA], 2020)

	Level 2	Level 3	Level 4
	Year 2	Year 4	Year 6
Sub element -Become confident, resilient, and adaptable	Undertake and persist with short tasks, within the limits of personal safety	Persist with tasks when faced with challenges and adapt their approach where first attempts are not successful	Devise strategies and formulate plans to assist in the completion of challenging tasks and the maintenance of personal safety
Student behaviour	Behaviour was agitated and distressed when a task could not be completed immediately or gives up immediately and disengages	James recognises that this task if difficult and makes an attempt then pauses and seeks another attempt to uses the materials and is successful in completing the task and make the catapult	James reviews the task carefully before making a careful attempt with all the necessary materials

Table 10.9 Teacher and student reflective questions

Reflection questions	Did the task need to be broken down into pieces in order for James to complete it? What happened to James' behaviour when he got frustrated?	Did James complete the task with help/without help? What happened to James' behaviour when he got frustrated? What did James need to be successful? How did James feel when he was successful?
Questions to ask student	What do you do when this does not work? How does this make you feel? Can you think what happens next? This must be frustrating for you but how can you keep going?	This looks tricky how are you going? What will you do next? Is there another way you can try? This must be frustrating for you but how can you keep going? Ok what will you do now and how can you complete this task?

10.7 Wigglebot Example Part 2

Part 2 of this example focuses on the General Capabilities. Figure 10.3 has been created as a blank template for students to describe a created Makerspace artefact (the centre circle of the figure), the content explored in that activity through the STEM learning areas (the second circle), and finally which Transversal Competencies are being demonstrated at each stage of the artefact's creation. Figures 10.4, 10.5 and

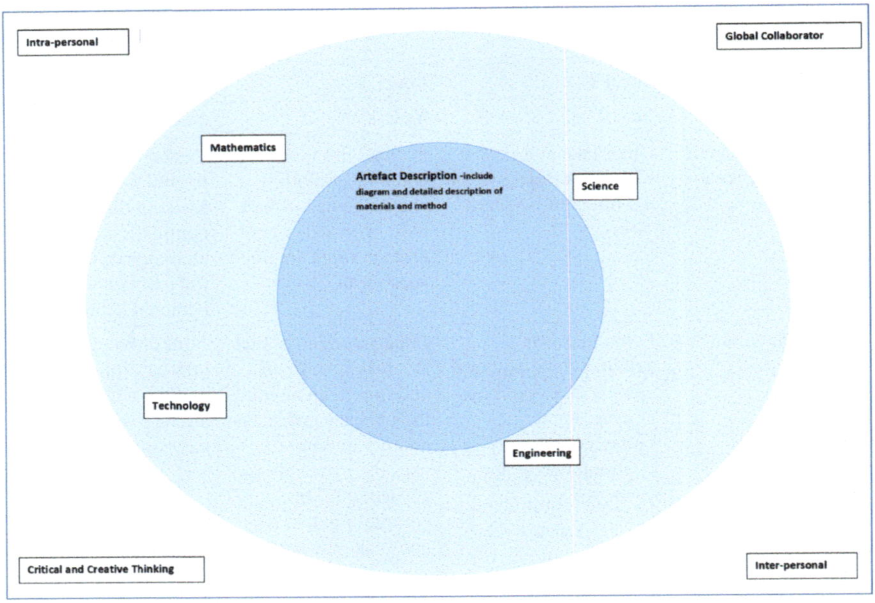

Fig. 10.3 Transversal competencies in the classroom

10.6 are three sample programs that illustrate how the relevant learning area and GC content descriptors can be documented for the Wigglebot activity. This approach can be applied to any other Makerspace activity.

10.8 Summary

How is STEM taught in the Australian Curriculum?
There is an aim for the STEM subjects to be taught holistically in the Australian Curriculum; however, they are listed as separate subjects that have their own strands, sub-strands, and content descriptors. Engineering is not identified as a specific subject but has elements which are subsumed by the *Science Inquiry* strand and the *Process and Production* strand of *Design and Technologies*.

What are the General Capabilities in the Australian Curriculum?
In the Australian Curriculum, there are seven General Capabilities:

- Literacy;
- Numeracy;
- Information, Communication, and Technology;
- Critical and Creative Thinking;
- Personal and Social Capability;

Inter-personal

- Making decisions: students are able to identify and consider potential influences that could impact their decision making processes

Critical and Creative Thinking

- Seek alternatives and put ideas into action: Students test and assess various processes before identifying and implementing the most effective solutions
- Reflect on processes: students are able to reflect and justify on the processes and choices they made during the method

Mathematics Content

- Able to solve problems that require comparison and investigation into determining equivalent lengths of the textas as the legs (ACMMG137)
- Representing and constructing simple pyramids with the Textas as the legs (ACMMG140)

Not Applicable

Design and Technology

Wigglebot- Year 6

Materials: Cup, peg, popstick, googly eyes, tape, scissors, 2xAA Batteries, motor, battery pack, 3x Textas

Method:

1. Put batteries into the battery pack
2. Attach battery pack to motor and tape into position on top of cup
3. Tape textas (with the lids facing down) evenly around the bottom of the cup
4. Tape the popstick to the peg, before attaching peg to the motor
5. Glue on googly eyes to outside of the cup
6. Turn on motor and watch Wigglebot rotate/woggle

Science Content

- Ability to create a complete electrical circuit that transfers energy (ACSSU097)
- Able to use scientific knowledge to determine the purpose and role of the electricity (ACSHE100).
- Able to pose questions that help to predict potential outcomes during the making process (ACSIS232)
- Able to reflect and identify potential improvements that could be implemented to future methods and design features to improve overall quality (ACSIS108)

- Investigating how the features of electrical circuits (eg. switches) can control the movement of the Wigglebot (ACTDEK020).
- - Ability to use rotations and translations to move the positioning and placement of battery pack and the textas legs (ACMMG142)

Engineering Content

Global Citizenship

- Exploring values: students are able to identify potential consequences of various actions during the making process eg. not wasting tape

Intra-personal

- Developing reflective practice: students thinking about their actions and responding to feedback from the teacher to help fulfil their goals and potential
- Becoming confident, resilient and adaptable: Students demonstrate abilities to persevere and adapt through challenging situations that may arise

Fig. 10.4 Makerspace activity demonstrating transversal competencies example 1

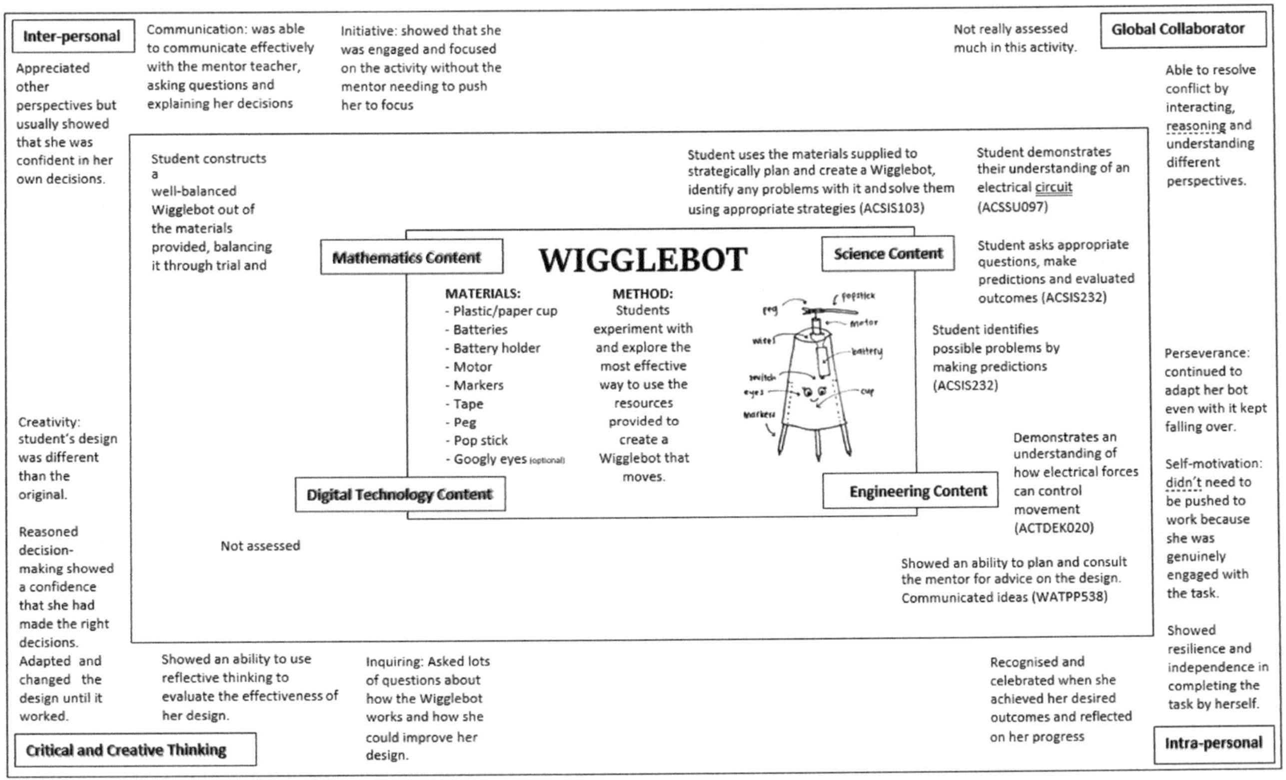

(Australian Curriculum, n.d), (UNESCO, 2019).

Fig. 10.5 Makerspace activity demonstrating transversal competencies example 2

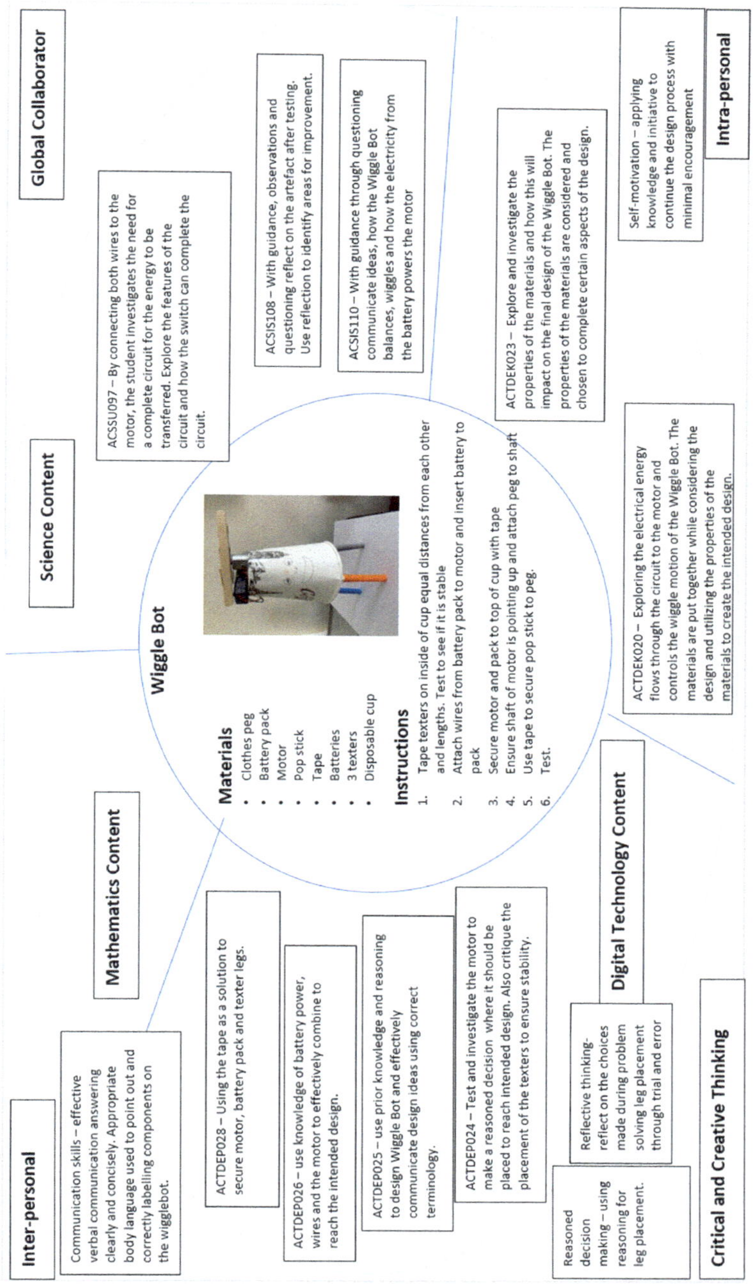

Fig. 10.6 Makerspace activity demonstrating transversal competencies example 3

- Intercultural Understanding; and
- Ethical Understanding.

Where are the Transversal Competencies situated in the Australian Curriculum?

Transversal Competencies are represented in the Australian Curriculum as the General Capabilities. Table 10.4 illustrates the relationship between the Transversal Competencies and the General Capabilities, and the common language used in both frameworks.

Where do Makerspaces 'fit' into the Australian Curriculum?

Makerspaces facilitate STEM-based activities as they usually comprise a strong science or technology component where students. Although not explicitly mentioned in Australian Curriculum documents, the natural link between Makerspaces and the Australian Curriculum is through a STEM program. Most Makerspace activities comprise a strong science or technology component where students are solving a problem or using the design process and are using their personalised Makerspaces to develop their STEM programs.

Where do Makerspaces 'fit' into the Australian Curriculum?

Makerspaces are not directly related to specific learning areas or General Capabilities; however, they can provide a physical space or conceptual framework where a range of learner-based activities can be undertaken. The non-subject-specific approach of a Makerspace broadens its potential to include more integrative activities. This allows the development and assessment of individual STEM subjects (Science, Technology, and Mathematics) as well as a range of General Capabilities.

Chapter 11
Future-Proofing Makerspaces

When the wind of change blows, some people build walls, others build windmills (Chinese proverb)

Keywords Fluid learning · Education 5.0 · Agency · Scalability · Virtual Makerspaces · Ecosystems · Future-proofing

Questions

- Why is it important to ensure the provision of Makerspaces now and in the future?
- What are the essential elements of a Makerspace that need to be maintained/retained?
- What will/should Makerspaces look like moving forward?
- How do we future-proof a Makerspace?

> The gap between the industry's needs and the scope and quality of competencies offered generally by education is revealing the increasing need to find new educational models to face the new era of the smart digital connected world (Tan, 2018).

R. B. Koul et al., *Teaching 21st Century Skills*,
https://doi.org/10.1007/978-981-16-4361-3_11

11.1 Past, Present, and Future

> By 2020 we could have well-trained animal employees, including ape chauffeurs.
> (RAND Corporation Long-Range Forecasting Study, 1968)

If we peer too far into the future, we may up with the ridiculous, as is the case with the above quote. This chapter examines this uncertainty and unpacks a number of ideas about the future of education and then considers how Makerspaces fit within this rapidly evolving paradigm. We review the educational past and the continually evolving models of education to make tentative suggestions for the future. We articulate how Makerspaces have been developed from the research of Papert and how the criteria of a Makerspace approach are more in line with the learning situations, pedagogies, and tools suggested in Education 4.0 (Salmon, 2019) than in current educational practices. Finally, we describe the elements of a Makerspace considered essential for new learning paradigms.

Institutionalised education was developed for the purpose of meeting the needs of the industrial revolution in the sixteenth century, when the first free public schools were created in the US and UK (Robinson, 2010). These school systems were based on the traditional view of academic ability at the time, and that people were either academic (versed in the classics and deductive reasoning) or non-academic (focused on developing required industrial skills). Schools have continued to value traditional academic knowledge and are still modelled on the image of industrialism hundreds of years after it was introduced. As Robinson points out, in current school settings, bells are still used and there is little choice about what is learnt. Learning is delivered in discrete siloed subjects and students move through their school career based on their age and not their level of development or interest (Robinson, 2010). He argues that this level of conformity and the continued focus on standardised testing is detrimental to the progress in education and that there needs to be significant systematic change. This notion is reinforced by the OECD who suggests that:

> Education must evolve to continue to deliver on its mission of supporting individuals to develop as persons, citizens and professionals. It must remain relevant to continue to shape our children's identity and integration into society. In a complex and quickly changing world, this might require the reorganisation of formal and informal learning environments, and reimagining education content and delivery. In an ageing world, these changes are likely to apply not just to basic education, but to lifelong learning as well (OECD, 2020).

The inability of education systems to keep abreast of societal change is highlighted by data that shows an estimated 65% of students entering primary schools will work in new jobs that are yet to be created and 50% of the technical knowledge that an undergraduate studies will be obsolete by the time they graduate (Makrides, 2019).

Masters (2015, 2021) notes that preparing students for the twenty-first century was an issue in 2015—and is still an issue in 2021—with little focus in regular classes on Transversal Competencies or General Capabilities. He argues that a focus on these capabilities prepares students for a future that is uncertain and where learners will need to be continually flexible, agile, and disposed to continual learning. Another challenge is the need for a more flexible learning arrangement that caters to individual needs, interests, and abilities.

11.2 The Evolution from Education 1.0 to Education 4.0

While there are still prevailing challenges in education, there needs to be ongoing development of teaching and learning practices. In recent years, these have been described in terms of numbered 'versions' which reflect the changing nature of education through a series of paradigm shifts. As we focus on the key elements of learner, curriculum, pedagogy, content, assessment, and purpose, comparisons can be made with their apparent importance in each iteration. In 2015 to 2021, there is still evidence of schooling embracing Education 1.0 and Education 2.0, especially in some developing countries. These schools are not yet equipped to engage students around twenty-first-century skills or competencies or to ensure they are agile and responsive to change.

Table 11.1 describes movement in more progressive schooling from fixed learning of key content to a more fluid and flexible learning that is connected to the needs of the students. There is movement from an *instructivist* to a *constructivist* approach, where instead of receiving, responding, and regurgitating information, students are connecting, collecting, and curating to create something new. In this scenario, learners are encouraged to become less passive and more active, either co-constructing their learning or using AI to respond agilely to situations. Even learning spaces which started with fixed front facing desks have moved towards a more flexible arrangement where students move, sit and stand (with or without desks) depending on the learning situation.

While Education 4.0 is still not at the forefront of educational dialogue and implementation, it is gaining traction as a topic for robust discussion when considering where education needs to focus into the future. In summary, Education 4.0 features the following:

- learner driven learning;
- learners as agents of their own learning;
- learners as partners/collaborators;
- seamless learning (not bound by weeks, semesters or locations);
- personalised learning pathways;
- multiple learning providers (dependent on the needs of the learner);
- accumulation of learning credentials to create a learner profile that is constantly updated as lifelong learning continues;
- constructive digitalisation; and
- convergence of AI and learning experiences (immersive technologies).

In terms of skill development, the World Economic Forum (2021) suggests that curriculum under Education 4.0 must cultivate these skills:

Table 11.1 Education evolution from 1.0 to 4.0 (based on Leapfrog Principle)

Attributes	Education 1.0	Education 2.0	Education 3.0	Education 4.0
Model	Download education	Open access education	Knowledge producing education	Innovation producing education
Curriculum	Fixed knowledge base Exams	Variety of knowledge Exams	Knowledge and skill based	Fluid learning skills and driven by innovation and learners needs
Approach and focus	Instructivist (3Rs) Respond, receive and regurgitate	Constructivist 4Cs Communicate, contribute, collaborate and co-creating	Connectivist 3Cs Connecting, collecting, and curating	Adaptive learning driven by AI Learning process based on real-time student profile Built through selective individual and team-driven embodiments in practice through innovation
Educator actions	Knowledge source	Facilitator teams with students	Learning designer of collaborative knowledge creation	Supported by an AI learning portal (super agile and responsive)
Student actions	Passive learners	Active learners	Co-developers and co-researchers Authors, driver and assessors of learning experiences	Self-governed autonomous Counsellors and AI co-develop education plans which constantly evolve Input of learning is the major source of technology evolution
Technology use	Internet age starting	E learning and collaboration Open source starting	Low-cost digital mobile Web driven technologies Use purposely for the selective production of knowledge	Personalised intelligent models, IOT Web driven e-learning
Space	Rows of desks chairs facing teachers desk at front	Groups desks and chairs seeing the teacher at front	Groups of seats but no teacher at the front	No teachers desk in the room students can be anywhere and there are no formal seating

- global citizenship;
- innovation and creativity;
- technology; and
- interpersonal skills.

These skills have been discussed in Chaps. 4, 5, 10. Teaching practices must also support the following types of learning experiences:

- personalised and self-paced learning;
- accessible and inclusive learning;
- inquiry-based learning;
- problem-based and collaborative learning; and
- lifelong and student-driven learning.

The Scenarios of Future of Schooling (OECD, 2020) suggests that many of the elements listed above would be difficult to facilitate within current educational programs. As discussed previously, with the majority of school systems still teaching siloed subjects in the secondary schools, and being driven by external exams, including A and O levels (UK), board exams (India) and ATAR (Australia), there is still an emphasis on traditional subjects with content-loaded exams used to enable matriculation. While the World Economic Forum offers examples of schools who have restructured learning in order to promote these shifts to curriculum and teaching practices, they are not necessarily realistic for the average school.

The OECD describes a number of possible scenarios for the future of schooling in the Back to the Future of Education: Four OECD Scenarios for Schooling (2021) document. The first is *extended schooling* where formal education expands to include international collaboration and individualised learning. The structures and processes of schooling remain. The second scenario is where education is *outsourced and privatised* with competing providers. This leads to the diversification of opportunities with technology as a key driver. The third scenario is where schools are *learning hubs* with a connection to the community through face-to-face and online methods. These flexible schooling arrangements would permit greater personalisation and community involvement. In this setting, teachers are considered nodes in wider networks of flexible expertise. Finally, the fourth scenario is that education can be *anywhere and anytime* with few distinctions between formal and informal education. This scenario would see schools dismantled and teachers become a mobile and active commodity. While all of these scenarios may seem far-fetched at present, there are elements of each scenario that are currently in practice.

> The horse is here to stay but the automobile is only a novelty – a fad.
>
> President of Michigan Savings Bank, warning Henry Ford's [inventor of the automobile] lawyer not to invest in the Ford Motor Company, 1903 (OECD, 2021).

11.3 Makerspaces and Education 4.0

Rapid technological innovation has left educators reluctant to engage deeply in a Makerspace program due to the perception that whatever they introduce may quickly become obsolete. Therefore, they are hesitant to invest time and resources into long-term projects. This perception is at odds with the reality that, as education has transitioned through 1.0 to 4.0, there appears to be a closer alignment with the underlying principles of a Makerspace as illustrated in Fig. 11.1.

We advocate that Makerspaces and the approaches suggested in this book are not transient but are rooted in the work of Piaget's and Papert's 'constructivism', 'constructionism', and 'play' theories that have been applied to learning pedagogies since the early 1950s. The constant re-emergence and referral to these philosophies show their durability. As versions 1.0 through to 4.0 of education are converging towards and aligning with the principles and practices on which Makerspace are based, it can be seen as the perfect place to provide students with agency while focusing on problem-based learning and collaboration.

Table 11.2 describes how Makerspaces can accommodate future learning to illustrate the natural alignment between Education 4.0 and Makerspaces.

To bring STEM Makerspaces from the fringes of STEM clubs, after school and incidental programs, changes are needed in the following areas:

- Assessment and exam focus. Students are regulated and bound by exams which reward remembering and regurgitation with little emphasis on skill development. Removal of current exams and standardised testing is important to encourage

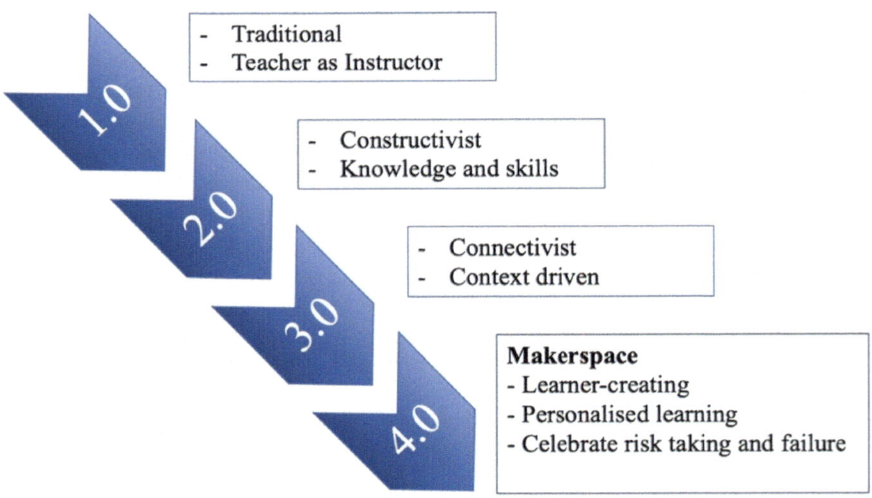

Fig. 11.1 Education 1.0 to education 4.0

Table 11.2 Future of learning corresponding to Makerspaces

Future of learning	Makerspaces
Learner driven learning Learners as agents of their own learning Personalised learning Accessible and inclusive learning	Community members (students) are able to pursue their own creation or a version of their own creation depending on the approach and outcome. Everyone can achieve dependent on their skills, their learning and their aims
Learners as partners and collaborators Problem-based and **collaborative learning** Interpersonal skills	Working together on projects which enables members to bring their expertise to the space and share or work together to learn new skills
Seamless learning—not bound by weeks, semesters or locations. Personalised learning pathways	Makerspace are usually project or task orientated so that learning is personalised and bounded by an individual's needs Spaces are unique and meet the needs of the students and are unbound by traditional desks and chairs
Constructive digitalisation	Creating and constructing either physically or through immersive technologies
Accumulation of learning credentials to create a learner profile that is constantly updated as lifelong learning continues	Community Makerspaces exist and are places where members continue to learn and develop their skills and micro-credentials are awarded to show accomplishment

personalised learning by removing a siloed approach and increasing student agency.

- Timetables and flexible learning opportunities. Currently regimented timetables and siloed subjects do not allow for the organic and integrated approach that is espoused by Education 4.0 and Makerspaces.

The vision of 4.0 may seem unachievable in the short-term; however, Makerspaces provide a gateway into exploring the possibilities of what Education 4.0 may look like within an existing school system. For example, teachers can experiment with new pedagogical approaches and assessment strategies in a format that does not have significant consequences to the schools existing program. Makerspace could also be considered a professional learning playground for staff to develop skills in problem-based and inquiry-based learning, digital technology competency, and classroom management practices.

11.4 Virtual Makerspaces

> A virtual environment where students and adults can create, build, and invent and where all the other creative, informal, educational self-directed learning passions can develop (Loertscher on Virtual Makerspaces, 2015)

A virtual Makerspace is a one-stop, web-based space where users may access digital tools for engaging in online maker-style activities. Like a physical Makerspace, they are focused on inquiry, critical thinking, and problem-solving skills. It may seem that creating a Makerspace is only possible in a physical room, but as learnt by the COVID-19 pandemic, being able to adapt to remote and hybrid learning is a necessity. Virtual Makerspaces are being designed by teachers and technical specialists to keep this type of creative learning possible for students. Virtual laboratory packages are also growing in popularity in high schools and middle schools. K-12 science teachers are able to use a simulation or game to capture students' interest to explore topics in an immersive environment.

Examples of virtual Makerspaces

Simulations and games can have great educational value. These tools allow students some control over the pacing and content they are exposed to, which translates to an opportunity for individualised learning that matches each student's unique needs and interests. Using games as a learning tool is effective because students are already exposed to playing games for entertainment. Some students, who would normally not be interested in learning science, may find science fascinating once they are introduced to educational games.

Minecraft can be considered as a virtual Makerspace, as discussed in Chap. 3 with the Elmtree Primary School example. Miss Claire organised a virtual Minecraft field trip for students to experience and research the type of Makerspace they wished to develop. Students created a real and virtual Makerspace where they explored Australian Native Bees and created bee hotels.

Labster represents the experience that students would go through when undertaking a laboratory investigation, from putting on their lab coats to looking down the microscope. Students then answer a series of questions and are provided with educative feedback if they make errors. Teachers also have an analytical dashboard to monitor and assess student engagement.

The Mystery of the Taiga River is a game-based science project that immerses students aged 10-14 in the Taiga River National Park where investigative reporters and scientists wished to find out what's behind the demise of the river. Students test water quality and talk with various stakeholders (loggers, farmers, fishermen) to restore the health of the river. With this simulation and others, students can travel to virtual places to play games that have educational value. In the simulation, students are able to create their own persona and communicate with classmates and teachers.

The Atlantis Remix Project is aimed at children ages 9–16 and has online and off-line learning activities for kids. More than 100,000 children have participated in the Quest Atlantis and Atlantis Remixed projects.

STEAM Craft Edu's Planeteers is an award-winning game that features a virtual Makerspace and open-sandbox experience for students to build and craft anything they can imagine. The game also features a 3D coding platform which encourages the development of STEAM focused skills.

We were going to include another example of a virtual Makerspace called **Workbench** which was available for 3.7 million card holders from Maryland Public Libraries. It was a Makerspace that provided access to activities and lessons, let users discuss and share projects, and upload their own ideas. Sadly, when we went to explore the site, it had closed down in 2020. This is an example of the potential impermanence of the virtual space, where a resource can be shut down at any time without notice.

11.5 Summary

In concluding this chapter, we return to the beginning of our investigations into the value of Makerspaces. Table 11.3 poses a series of guided questions designed to assist with a functional Makerspace proposal for your school administration, colleagues, parents, and students.

Why is it important to ensure the provision of Makerspaces now and in the future?

As education has moved from 1.0 to 4.0, elements of Makerspaces have been in present increasingly prevalent as you move through the versions. If this trend continues, then one can extrapolate that the essential elements of a Makerspace will be more important than ever to prepare for the future world of work. Makerspace

Table 11.3 Value of Makerspaces

Why	Why will this Makerspace be created in the school?	The rationale statement that will be presented to the administration
How	How will success be measured? Assessment How will impact be measured? Reporting	Success indicators for impact Success indicators for learning
What	What is the purpose of Makerspace? What is the content to be learnt? What are the skills to be learnt?	
Who	Who will be involved in Makerspace? Who are the students? Who are the support staff Who else will be using or sharing the space? Who will create the space	
Where	Where will the real or actual Makerspace be situated? What will this space also used for? Where will the virtual space be situated	
When	When would the classes be scheduled, or would this be an after-school activity?	

fundamentals are being recognised and aligned to the evolving characteristics that are considered essential in twenty-first-century learning.

What are the essential elements of a Makerspace that need to be maintained/retained moving forward?

- Fluid, adaptable, and open;
- learner centred;
- community based;
- reflects current technologies;
- encourages risk-taking and exploration, and;
- nurtures creativity.

What will / should Makerspaces look like moving forward?
It is difficult to predict what Makerspaces will look like into the future; however, the aspects of a Makerspace that are aligned closely with Education 4.0 are;

- learners as agents of their own learning;
- learners as partners/collaborators;
- seamless or fluid learning—not bound by weeks, semesters or locations;
- personalised learning pathways that involve higher-order thinking;
- accumulation of learning credentials to create a learner profile that is constantly updated; as lifelong learning continues, and;
- constructive digitalisation such as AI into the learning experiences.

How do we future-proof a Makerspace?
A Makerspaces lifespan is determined by its adaptability to new technologies, ideas, and pedagogical approaches that support the development of essential twenty-first-century skills. Maintaining a safe and supportive environment will ensure its continuity, as will:

- someone to champion the program;
- a strong foundation, i.e clearly identified purpose and goals;
- regular review of Makerspace to ensure it stays relevant to changing contexts and situations;
- multiple interest groups;
- something for everyone, and;
- authentic activities.

Appendix

Makerspace Rationale

Complete the questions and then remove them to leave the rationale on the right-hand side ready to be presented to the administration.

		Makerspace Rationale
Why	Why will this Makerspace be created in the school?	The rationale statement that will be presented to the administration
How	How will success be measured? Assessment How will impact be measured? Reporting	Success indicators for impact Success indicators for learning
What	What is the purpose of Makerspace? What is the content to be learnt? What are the skills to be learnt?	
Who	Who will be involved in Makerspace? Who are the students? Who are the support staff Who else will be using or sharing the space? Who will create the space	
Where	Where will the real or actual Makerspace be situated? What will this space also used for? Where will the virtual space be situated	
When	When would the classes be scheduled, or would this be an after-school activity?	

References

Aayog, N. (2017). *Application for Atal Tinkering Laboratories*. http://niti.gov.in/content/atal-tinkering-laboratories

Adam, S. (2004). Using learning outcomes—A consideration of the nature, role, application and implications for European education of employing 'learning outcomes' at the local, national and international levels. Presented at the United Kingdom Bologna Seminar.

Alber, R. (2017, June 19). Deeper learning: A collaborative classroom is key. Edutopia. https://www.edutopia.org/blog/deeper-learning-collaboration-key-rebecca-alber

American Society for Quality (ASQ) (n.d.) What is problem-solving?. https://asq.org/quality-resources/problem-solving

Anderson-Levitt, K. (2017). Global flows of competence-based approaches in primary and secondary education. *Cahiers De La Recherche Sur L'éducation Et Les Savoirs, 16*, 47–72.

Andrews, K. (2015). Jobs of the future: The known unknown. *Australasian Science, 36*(7), 38.

Assessment and Teaching of 21st Century Skills. (2012). ATCS Project. http://www.atc21s.org

Attewell, J. et al. (forthcoming). Guidelines for makerspaces in schools. European Schoolnet.

Australian Curriculum (n.d.) Personal and social capability learning continuum. https://www.australiancurriculum.edu.au/media/1078/general-capabilities-personal-and-social-capability-learning-continuum.pdf

Australian Curriculum. (2020). General capabilities: Critical and creative thinking https://australiancurriculum.edu.au/f-10-curriculum/general-capabilities/critical-and-creative-thinking/

Australian Curriculum, Assessment and Reporting Authority [ACARA]. (2018). Australian curriculum general capabilities. https://www.australiancurriculum.edu.au/f-10-curriculum/general-capabilities/

Australian Curriculum, Assessment and Reporting Authority [ACARA]. (2016). ACARA STEM connections project report. https://www.australiancurriculum.edu.au/media/3220/stem-connections-report.pdf

Australian Government. (2018). Inspiring all Australians in science, technology, engineering and mathematics. https://www.innovation.gov.au/page/inspiring-nation-scientists. Australian Industry Group.

Australian Government, Department of Education (n.d.) Support for Science, Technology, Engineering and Mathematics (STEM). v.au/support-science-technology-engineering-and-mathematics.

Australian Industry Group. (2015). Progressing STEM skills in Australia [Melbourne]. Available http://cdn.aigroup.com.au/Reports/2015/14571_STEM_Skills_Report_Final_-.pdf

Awanga, H., & Zawawi, D. (nd) Improving a communication skill through the learning approach towards the environment of engineering. https://doi.org/10.1016/j.sbspro.2015.06.241

R. B. Koul et al., *Teaching 21st Century Skills*,
https://doi.org/10.1007/978-981-16-4361-3

Bailin, S. (1987). Critical and creative thinking. Informal Logic (9)1 Winter, University of Manitoba.

Baker, M., Rudd, R. D., & Pomeroy, C. S. (2000). Creativity of low resource farmers in the Dominican Republic. In *Proceedings of the Association for International Agricultural and Extension Education's Annual Conference.*

Barron, B., & Darling-Hammond, L. (2010). Prospects and challenges for inquiry-based approaches to learning. In H. Dumant, D. Istance, & F. Benavides (Eds.), *The nature of learning: Using research to inspire practice, education research and innovation*, OECD Publishing Paris. https://doi.org/10.1787/9789264086487-en

Barry, T. (2015) Library as place, make the space: Makerspaces as community development. Australian Library and Information Association. http://information-online.alia.org.au/content/library-place-make-space-makerspaces-community-development

Becker, K., & Parker, K. (2011, July-Sept). Effects of integrative approaches among STEM subjects on students' learning. *Journal of STEM Education, 12.*

Beede, D., Julian, T.A., & Langdon, D. (2011). Women in STEM: A gender gap to innovation. *SSRN Electronic Journal*. https://doi.org/10.2139/ssrn.1964782

Bennett, B. (2013). Teaching and learning strategies. School Drug Education and Road Aware. https://www.sdera.wa.edu.au/media/1268/teaching-and-learning-strategies-yr-6.pdf

Beyer, B. K. (1987). *Practical strategies for the teaching of thinking.* Allyn and Bacon Inc.

Beyond Blue Ltd. (2017). Building resilience in children aged 0–12: A practice guide https://resources.beyondblue.org.au/prism/file?token=BL/1810_A

Bieraugel, M., & Neill, S. (2017). Ascending Bloom's pyramid: Fostering student creativity and innovation in academic library spaces. *College and Research Libraries, 78*(1), 35–52. https://doi.org/10.5860/crl.v78i1.16566

Bird, D., & Lindsey, B. (2019) Hacking Maslow: Assessment, scale and survival of a statewide community college makerspace network in California. ISAM 2019 Paper No 14, 10/17/19. https://cccmaker.com/wp-content/uploads/2019/10/ISAM-2019-Presentation-Hacking-Maslow.pdf

Blackley, S., & Howell, J. (2015). A STEM narrative: 15 years in the making. *Australian Journal of Teacher Education, 40*(7), 102–112. https://doi.org/10.14221/ajte.2015v40n7.8

Blackley, S., Rahmawati, Y., Fitriani, E., Sheffield, R., & Koul, R. (2018). Using a makerspace approach to engage Indonesian primary students with STEM. *Issues in Educational Research, 28*(1), 18–42. http://www.iier.org.au/iier28/blackley.pdf

Blikstein, P., & Krannich, D. (2013). The makers' movement and FabLabs in education: Experiences, technologies, and research. In *Proceedings of the 12th International Conference on Interaction Design and Children*, (pp. 613–616).

Blikstein, P., Martinez, S.L., & Pang, H.A. (2016). Meaningful making: Projects and inspirations for Fab Labs+ makerspaces. Constructing Modern Knowledge Press.

Blikstein, P. (2018). Maker movement in education: History and prospects. In M. J. de Vries (Ed.), *Handbook of technology education*, Springer International Handbooks of Education. 11.1007/978–3–3 19–44687–5_33

Bloom, B. S., Engelhard, M. D., Furst, E. J., Hill, W., & Krathwohl, D. R. (1956). *Taxonomy of educational objectives: The classification of educational goals. Handbook I: Cognitive domain.* David McKay Company.

Blumschein, P. (2012). Intentional learning. In N. M. Seel (Ed.), *Encyclopedia of the sciences of learning* (pp. 1600–1601). Springer.

Blum-Ross, A., Kumpulainen, K., & Marsh, J. (forthcoming) Enhancing digital literacy and creativity: Makerspaces in the early years. Routledge. https://www.crcpress.com/Enhancing-DigitalLiteracy-and-Creativity-Makerspaces-in-the-Early-Years/Blum-Ross-KumpulainenMarsh/p/book/9780367197889

Bower, M., Stevenson, M., Falloon, G., Forbes, A., & Hatzigianni, M. (2018) Makerspaces in primary school settings—Advancing 21st century and STEM capabilities using 3D Design and 3D printing. Macquarie University. https://www.makersempire.com/wp-content/uploads/2018/09/Makerspaces-in-Primary-School-Settings-Full-Report.pdf

Bowler, L. (2014). Creativity through "maker" experiences and design thinking in the education of librarians. *Knowledge Quest, 42*(5), 58–61.

Bowler, L., & Champagne, R. (2016). Mindful makers: Question prompts to help guide young peoples' critical technical practices in maker spaces in libraries, museums, and community-based youth organizations. *Library and Information Science Research, 38*(2), 117–212. https://doi.org/10.1016/j.lisr.2016.04.006

Bozik, M. (1987, November 5–8). Critical thinking through creative thinking. *Paper presented at the annual meeting of the speech communication association* (73rd). PUB TYPE Speeches/Conference Papers (150). https://files.eric.ed.gov/fulltext/ED290173.pdf

Bocconi, S., Chioccariello, A., Dettori, G., Ferrari, A., Engelhardt, K., Kampylis, P., & Punie, Y. (2016). Developing computational thinking: Approaches and orientations in K-12 education. In G. Veletsianos (Eds.), *EdMedia 2016 Proceeding*. Association for the Advancement of Computing in Education (AACE) (pp. 13–18).

Brown, T. (2020). The social administration of mathematics subject knowledge through teacher education. In *A contemporary theory of mathematics education research* (pp. 49–72). https://doi.org/10.1007/978-3-030-55100-1_5

Burden, P. R., & Byrd, D. M. (1994). *Methods for effective teaching*. Allyn and Bacon Inc.

Burke, J. (2014). Makerspaces: A practical guide for librarians. Rowman & Littlefield https://dictionary.cambridge.org/dictionary/english/collaboration

Bybee, R. W. (2014). The BSCS 5E instructional model: Personal reflections and contemporary implications. *Science and Children, 51*(8), 10–13.

Cachero, P., Edgett, K., & Tsur, S. (2017) Jugaad: An untranslatable word for winging it. BBC. https://www.bbc.com/culture/article/20171213-jugaad-an-untranslatable-word-for-winging-it

Caena, F. (2019). Developing a European framework for the personal, social and learning to learn key competence (LifEComp). *Literature Review and Analysis of Frameworks. Publications Office of the European Union*https://doi.org/10.2760/172528

Cahill, H., Beadle, S., Farrelly, A., Forster, R., & Smith, K. (2014). Building resilience in children and young people: A literature review for the Department of Education and Early Childhood Development (DEECD). Department of Education, Victoria.

Caprile, M., Palmén, R., Sanz, P., & Dente, G. (2015). Encouraging STEM studies for the labour market. http://www.europarl.europa.eu/RegData/etudes/STUD/2015/542199/IPOL_STU(2015)542199_EN.pdf

Cavalcanti, G. (2013). *Is it a Hackerspace, Makerspace, TechShop, or FabLab? Make Magazine*. https://makezine.com/2013/05/22/the-difference-between-hackerspacesmakerspaces-techshops-and-fablabs/

Cedefop. (2016). Application of learning outcomes approaches across Europe. Cedefop reference series; No 105. Publications Office of the European Union website. https://www.cedefop.europa.eu/en/publications-and-resources/publications/3074

Cedefop. (2017). Defining, writing and applying learning outcomes—A European Handbook. Publications Office of the European Union. https://www.cedefop.europa.eu/en/publications-andresources/publications/4156

Centre for Curriculum Redesign. (2015). Skills for the 21st century: What should students learn. https://curriculumredesign.org/wp-content/uploads/CCR-Skills_FINAL_June2015.pdf

Children's Museum of Pittsburgh and the Institute of Museum and Libraries. (2017). *Making + Learning in museums and libraries: A practitioner's guide and framework*. The Children's Museum of Pittsburgh and the Institute of Museum and Library Services. https://makingandlearning.squarespace.com/publication/41

Chubb, I. (2015). Aspiring to something magnificent. Address to the National Press Club, European Union. Horizon 2020 Work Programme. 2014–2015.

Clapp, E. P., Ross, J., Ryan, J. O., Tishman, S., Berger, R., & Donner, W. (2017). *Maker-centered learning: Empowering young people to shape their worlds*. Jossey Bass.

Claro, M., Preiss, D. D., San Martín, E., Jara, I., Hinostroza, J. E., Valenzuela, S., et al. (2012). Assessment of 21st century ICT skills in Chile: Test design and results from high school level students. *Computers and Education, 59*(3), 1042–1053. https://doi.org/10.1016/j.compedu.2012.04.004

Clements, D. H., Vinh, M., Lim, C. I., & Saram, J. (2020). STEM for inclusive excellence and equity. *Early Education and Development, 32*(3), 1–24. https://doi.org/10.1080/10409289.1755776

Colagrossi, M. (2018). 10 reasons why Finland's education system is the best. https://bigthink.com/mike-colagrossi/no-standardized-tests-no-private-schools-no-stress-10-reasons-why-finlands-education-system-in-the-best-in-the-world

Connor, K., & Davidson, J. (2003). Development of a new resilience scale: The Connor-Davidson Resilience Scale (CD-RISC). *Depression and Anxiety8*, 76–82. https://static1.squarespace.com/static/51abe64ee4b0a1344208e98a/t/51d3ce77e4b001d5c13a7e87/1372835447995/CD-RISC+%28Connor+%26+Davidson%2C+2003%29.pdf

Cooper, A. (2016). The joy of making: This stem-based movement spurs critical thinking and creativity. *Scholastic Teacher, 125*(5), 24. (Rep.). http://www.nysci.org/media/file/DMP-Report-2012.pdf

Costa, A., & Kallick, B. (2009). *Habits of mind across the curriculum.* Association for Supervision and Curriculum Development.

Cox, J. (2018). Teaching strategies to promote critical thinking. http://www.teachhub.com/teaching-strategies-promote-critical-thinking

Crane, L. D. (1983). Unlocking the brain's two powerful learning systems. *Human Intelligence Newsletter, 4*(4), 7.

Cresswell, J. W., Plano, C., & Vicki, L. (2007). *Designing and conducting mixed methods research.* Sage.

Crichton, S., & Carter, D. (2015). Maker day toolkit ver. 2. Innovative Learning Centre. https://issuu.com/ubcedo/docs/makerdaytoolkitver2revisemay31e

Crichton, S., & Carter, D. (2017). Taking making into classrooms: A toolkit for fostering creativity and imagination. Ministry of Education, Province of British Columbia. https://www.researchgate.net/publication/314205990_Taking_Making_Into_Classrooms_A_Toolkit_for_Fostering_Creativity_and_Imagination

Crockett, L., Jukes, I., & Churches, A. (2011) In *Literacy is not enough: 21st century fluencies for the digital age.* SAGE Publications.

de Boer, J. (2015). The business case of FryskLab, Europe's first mobile library FabLab. *Library Hi Tech, 33*(4), 505–518.

Daniel, O.B., & Redlich, T. (2015). Fablab research report 2015. https://doi.org/10.13140/RG.2.2.16966.06723

Declicin. (2012). Evaluation form of skills and knowledge acquired. EU Lifelong learning programme project report. http://www.asceps.org/makingprojects/declicin/wpcontent/uploads/Evaluation-of-skills-knowledge-acquired-in-Declicin.pdf

Declicin .(2013). Project write-up on the participation of women in the project activities. EU Lifelong learning programme project report. http://www.asceps.org/makingprojects/declicin/wp-content/uploads/Participation-of-females-inDeclicIn.pdf

Dewey, J. (2012). *Democracy and education.* Dover Publications.

DiyLab. (2016). D. 1.6. Report on digital competence in schools: Spain, Finland and the Czech Republic. EU lifelong learning programme project report. *Do it yourself in education: Expanding digital competence to foster student agency and collaborative* (71400ALLPA4A3541A4AESAKA1MP). http://diylab.eu/docs/D1.6_Report_on_Digital_Competence_in_Schools.pdf

Dougherty, D. (2012). Learning by making. https://slate.com/technology/2012/06/maker-faire-and-science-education-american-kids-should-be-building-rockets-and-robots-not-taking-standardized-tests.html

Dougherty, D. (2016). Free to make: How the maker movement is changing our schools, our jobs, and our minds. North Atlantic Books.

Duckworth, A. (2017). Grit: The power of passion and perseverance. Random House.

Dunmire, C. (2004). The four steps of creativity. (February 10, 2006). http://www.creativity.portal.com/main/privacy.html

Dwek, C. (2015). A growth mindset: The joy and power of reading. https://www.ahschools.us/cms/lib08/MN01909485/Centricity/Domain/4185/The%20Joy%20and%20Power%20of%20Reading.pdf#page=64

The Engineering Design Process (n.d.). https://www.eie.org/overview/engineering-design-process

Ennis, R. H. (1985). Goals for a critical thinking curriculum. In *Developing minds: A resource book for teaching thinking*. Association for Supervisions and Curriculum Development (ASCD) (pp. 54–57).

Eurydice. (2019). Digital education at school in Europe. Luxembourg: publications office of the European Union, Education, Audiovisual and Culture Executive Agency. https://eacea.ec.europa.eu/national-policies/eurydice/content/digital-education-school-europe_en

Europe: Department of Science and Technology. (2013). Research and development statistics at a glance 2011–12. http://www.nstmis-dst.org/pdf/finalrndstatisticsataglance2011121.pdf

European Commission. (2018). ATS202 Assessment of Transversal Skills. http://www.ats2020.eu.

European Commission. (2018). Commission staff working document accompanying the document proposal for a council recommendation on key competences for lifelong learning (No. SWD/2018/014 final-2018/08 (NLE)). https://eur-lex.europa.eu/legalcontent/EN/TXT/HTML/?uri=CELEX:52018SC0014&from=ENFabLab (2012)

European Regional Development Fund (n.d.) Urban manufacturing: Collaborative maker space case study summaries https://www.interregeurope.eu/fileadmin/user_upload/tx_tevprojects/library/file_1556879611.pdf

Evans, J. (2017). Making critical thinking real with digital content. https://tomorrow.org/speakup/MakingCriticalThinkingReal_CUE2017.html

Fasso, W., & Knight, B. A. (2019). Identity development in school makerspaces: Intentional design. *International Journal of Technology and Design Education, 14*(2), 80–98. https://doi.org/10.1007/s10798-019-09501-z

Finnish National Agency for Education. (2016). National Core Curriculum for Basic Education 2014. ISBN 9789521362590. https://verkkokauppa.oph.fi/EN/page/product/national-core-curriculum-forbasic-education-2014/2453039

Fleming, L. (2015). In *Worlds of making: Best practices for establishing a makerspace for your school*. Corwin Press. https://www.goodreads.com/work/best_book/44720004-worlds-ofmaking-best-practices-for-establishing-a-makerspace-for-your

Forman, E. A., & Cazden, C. B. (1986) Exploring Vygotskian perspectives in education: The cognitive value of peer interaction. In J. V. Wertsch (Ed.), *Culture, communication, and cognition: Vygotskian perspectives*. Cambridge University Press.

Foundation for Young Australians (FYA). (2019). Investing in young Australians: Fact sheet, https://www.fya.org.au/wp-content/uploads/2019/01/FYA_Campaign_National-FS_07.pdf

Fraillon, J., et al. (2014). Preparing for life in a digital age: The IEA international computer and information literacy study international report. https://research.acer.edu.au/ict_literacy/8

Framework for 21st-century learning definitions. (2019). http://www.battelleforkids.org/networks/p

Freeman, A., Adams Becker, S., Cummins, M., Davis, A., & Hall Giesinger, C. (2017). NMC/CoSN Horizon Report: 2017 K-12. The New Media Consortium.

Freeman, S., Eddy, S. L., McDonough, M., Smith, M. K., Okoroafor, N., Jordt, H., & Wenderoth, M. P. (2014). Active learning increases student performance in science, engineering, and mathematics. *Proceedings of the National Academy of Sciences, 111*(23), 8410–8415.

Frey, B., et al. (2006). Measuring collaboration among grant partners. *American Journal of Evaluation, 27*(3), 383–392.

Future Learn. (2021). The future of learning report. https://cdn-wordpress-info.futurelearn.com/info/wp-content/uploads/The-Future-of-Learning-Report-2021.pdf

Gale, N., Heath, G., Cameron, E., et al. (2013). Using the framework method for the analysis of qualitative data in multidisciplinary health research. *BMC Medical Research Methodology., 13*(1), 117.

Gao, Y., Borui, L., Wei, W., Wenyao, X., Chi, Z., & Zhanpeng, J. (2018). Watching and safeguarding your 3D printer: Online process monitoring against cyber-physical Attacks. *Proceedings of the ACM on Interactive, Mobile, Wearable and Ubiquitous Technologies, 2*, 1–27. https://doi.org/10.1145/3264918

Gazette of India. (2010). Right to education. Ministry of Human Resource Development. http://educationportal.mp.gov.in/rte/public/rte_act_2009.pdf

Giannakos, M., Divitini, M., & Iversen, O. S. (2017). Introduction for the special issue on 'maker technologies to foster engagement and creativity in learning.' *Entertainment Computing, 18*(143), 143–144. https://doi.org/10.1016/j.entcom.2016.11.001

Ginsburg, K. (2014). Building resilience in children and teens: Giving kids roots and wings (3rd ed.). American Academy of Pediatrics.

Ginsburg, K. (n.d) The 7 Cs: The essential building blocks of resilience. Fostering Resilience. http://www.fosteringresilience.com/7cs.php

Gomes, P, (2016), Stanford FabLearn's Paulo Blikstein on the efficacy of maker ed: It's about process, not products—EdSurge News. EdSurge. https://www.edsurge.com/news/2016-05-26-stanford-fablearn-s-paulo-blikstein-on-the-efficacyof-maker-ed-it-s-about-process-not-products42

Gonski, D., Boston, K., Arcus, T., O'Brien, L., Gould, V., Perry, L., Johnson, W., & Roberts, M. (2018). Through growth to achievement: report of the review to achieve educational excellence in Australian Schools, March 2018. Commonwealth of Australia.

Gonzales, N. A., Germán, M., Kim, S. Y., George, P., Fabrett, F. C., Millsap, R., & Dumka, L. E. (2008). Mexican American adolescents' cultural orientation, externalizing behavior and academic engagement: The role of traditional cultural values. *American Journal of Community Psychology., 41*(1–2), 151–164.

Goodlad, J., Klein, F., & Tye, K. (1979). The domains of curriculum and their study. In J. I. Goodlad & Associates et al. (Eds.), Curriculum inquiry: The study of curriculum practice (pp. 3–76). McGraw Hill.

Government of India. (2020). National Education Policy-2020. Ministry of Human Resource Development. https://www.education.gov.in/sites/upload_files/mhrd/files/NEP_Final_English_0.pdf

Government of India. (2014). Learning outcomes at the elementary stage. National Council of Educational Research and Training (NCERT).

https://ncert.nic.in/pdf/publication/otherpublications/tilops101.pdf

Government of India. (2013). Science, Technology and Innovation Policy. http://www.dst.gov.in/sites/default/files/STI%20Policy%202013-English.pdf

Graves, C. (2014). Teen experts guide makerspace makeover. *Knowledge Quest, 42*(4), 8–13.

Griffin, P., & Care, E. (Eds.). (2015). *Assessment and teaching of 21st century skills*. Springer. https://doi.org/10.1007/978-94-017-9395-7

Guerriero, S. (2017). Pedagogical knowledge and the changing nature of the teaching profession. OECD Publishing. http://www.oecd.org/education/pedagogical-knowledge-and-the-changingnature-of-the-teaching-profession-9789264270695-en.htm

Gutman, L. M., & Schoon, I. (2013). In *The impact of non-cognitive skills on outcomes for young people: Literature review*. Institute of Education.

https://educationendowmentfoundation.org.uk/public/files/Publications/EEF_Lit_Review_Non-CognitiveSkills.pdf

Gutsche, B. (2013). Make connect learn. https://www.webjunction.org/news/webjunction/Make-Connect-Learn.html

Hackling, M., Murcia, K., West, J., & Anderson, K. (2014). Optimising STEM education in WA schools part 1: Summary report. Edith Cowan Institute for Education Research.

Halpern, D. F. (1996). *Thought and knowledge: An introduction to critical thinking*. Lawrence Erlbaum Associates, Publishers.

Halverson, E. R., & Sheridan, K. M. (2014). The maker movement in education. *Harvard Educational Review, 84*(4), 495–504. https://doi.org/10.17763/haer.84.4.34j1g68140382063

Hart, C. (2016). Library makerspaces: Revolution or evolution? School catalogue information service *97* Term 2. https://www.scisdata.com/connections/issue-97/library-makerspaces-revolution-or-evolution/

Hipkins, R., Boyd, S., & Joyce, C. (2005). Documenting learning of the key competencies: What are the issues? A discussion paper. New Zealand Council for Educational Research (pp. 22–25).

Huiying, C., Xiaoqing, G., & Lung-Hsiang, W. (2017). An investigation of twenty-first century learners' competencis in China. Asia Pacific Education Review. Springer Nature

Hunter, J. L. (2017). STEM education in primary schools will fall flat unless serious issues are addressed. https://theconversation.com/stem-education-in-primary-schools-will-fall-flat-unless-serious-issues-are-addressed-88017

Hurtes, K., & Allen, L. R. (2001). Measuring resiliency in youth: The resiliency attitudes and skills profile. *Therapeutic Recreation Journal 35*(4), 333–347. https://static1.squarespace.com/static/51abe64ee4b0a1344208e98a/t/51d3cf15e4b0fc82c7b8ebd5/1372835605044/RASP+%28Hurtes+%26+Allen%2C+2001%29.pdf

Institute of Museum and Library Services. (2009). Museums, libraries, and 21st-century skills. https://www.imls.gov/sites/default/files/publications/documents/21stcenturyskills.pdf

International Journal of STEM Education. (2020). *5*(1), 1–19. https://doi.org/10.1186/s40594-018-0129-0

International Society for Technology in Education (ISTE). (n.d) ISTE Standards for Students. https://www.iste.org/standards/for-students

Johnston, M. P. (2018). Supporting STEM education: Needs assessment of south eastern rural teacher librarians. *School Libraries Worldwide, 24*(2), 62. https://doi.org/10.14265.24.2.005

Joshi, V. (2019). In *Science education in India*. Springer. https://doi.org/10.1007/978-981-13-9593-2_11

Joynes, C., Rossignoli, S., & Fenyiwa Amonoo-Kuofi, E. (2019). 21st century skills: evidence of issues in definition, demand and delivery for development contexts (K4D Helpdesk report). Institute of Development Studies.

Karyotaki, M., & Drigas, A. (2016). latest trends in problem solving assessment. https://doi.org/10.3991/ijes.v4i2.5800

Kennedy, T. J., & Odell, M. R. I. (2014). Engaging students in S.T.E.M. *Education Science Education International, 25*(3), 246–258. http://files.eric.ed.gov/fulltext/EJ1044508.pdf

Kitagawa, L., Pomba, E., & Davis, T. (2018). Plastic pollution to solution. *Science and Children, 55*(7), 38–45.

Koul, R., Verma, G., & Nargund-Joshi, V. (2019). Science Education in India(1). Springer. https://doi.org/10.1007/978-981-13-9593-2

Krulick, S., & Rudnick, J. A. (1988). Problem solving: A handbook for elementary school teachers. Allyn & Bacon. http://files.eric.ed.gov/fulltext/ED301459.pdf

Kurti, R. S., Kurti, D. L., & Fleming, L. (2014). The philosophy of educational makerspaces. Part 1 of Making an Educational Makerspace. *Teacher Librarian, 41*(5), 8–11.

Kurup, P. M., Li, X., & Powell, G. et al. (2019). Building future primary teachers' capacity in STEM: based on a platform of beliefs, understandings and intentions. *IJ STEM Ed 6*, 10. https://doi.org/10.1186/s40594-019-0164-5

Lahey, J. (2013). Why parents need to let their children fail. https://www.theatlantic.com/national/archive/2013/01/why-parents-need-to-let-their-children-fail/272603/

Larmand, A. (2019). Eduporium Weekly: Ideas for the values in your makerspaces. Eduporium. https://www.eduporium.com/store/blog/eduporium-weekly-ideas-for-the-values-in-your-makerspace/

Leopold, T. A., Ratcheva, V., & Zahidi, S. (2016). The future of jobs and skills. https://reports.weforum.org/future-of-jobs-2016/chapter-1-the-future-of-jobs-and-skills/

Li, Y., Wang, K., Xiao, Y., & Froyd, J. E. (2020). Research and trends in STEM education: A systematic review of journal publications. *International Journal of STEM Education, 7*(1), 11. https://doi.org/10.1186/s40594-020-00207-6

Lincoln, Y. S., & Guba, E. G. (1985). *Naturalistic inquiry*. Sage Publications.

Liu, L., Hao, J., von Davier, A. A., Kyllonen, P., & Zapata-Rivera, J. (2016). A tough nut to crack: measuring collaborative problem solving. In Y. Rosen, S. Ferrara, & M. Musharraf (Eds.), *Handbook of research on technology tools for real-world skill development* (pp. 344–359). IGI Global. https://doi.org/10.4018/978-1-4666-9441-5.ch013

Lock, J., Redmond, P., Orwin, L., Powell, A., Becker, S., Hollohan, P., & Johnson, C. (2020). Bridging distance: Practical and pedagogical implications of virtual makerspaces. *Journal of Computer Assisted Learning*. https://doi.org/10.1111/jcal.12452

Locke, J. Y., Campbell, M. A., & Kavanagh, D. (2012). Can a parent do too much for their child? An examination by parenting professionals of the concept of overparenting. *Journal of Psychologists and Counsellors in Schools, 22*(2), 249–265.

London Hackspace Ltd. Copyright. (2016). london.hackspace.org.uk

Lou, N. (2016). Rise of the makerspace. *Popular Science, 288*(2), 88.

Lofti, N. G., & Bahgat, D. (2019). Bio-inspired design system for the egyptian market: A short-term project case study: Towards a new agenda for design in the mediterranean region. *PAD Pages on Arts and Design, 16*, 3–19.

Luitel, B. (2020). Different pedagogical approaches integrated in STEAM education programs. https://www.youtube.com/watch?v=YOOEkNeJ-yg&fbclid=IwAR1eGpFdJM3akCXLdnNShwAbdQkCcV6CwWOEyYsHfZbC510ey4Fmuta5L70

MakEY Project. (2017). University of Sheffield website. https://researchportal.helsinki.fi/en/publications/makerspaces-in-the-early-years-a-literature-review

Makrides, G. (2019). The evolution of education from education 1.0 to education 4.0: Is it an evolution or a revolution? In *Conference of European Association of Erasmus Coordinators*. Beer Sheva, Israel, 4 March 2019.

Manninen, J. (2017). Empirical and genealogical analysis of non-vocational adult education in Europe. *International Review of Education, 63*, 319–340. https://doi.org/10.1007/s11159-017-9638-1

Mansbach, J. (2017). Using technology to develop students' critical thinking skills. https://dl.sps.northwestern.edu/blog/2015/09/using-technology-to-develop-students-critical-thinking-skills/

Marginson, S., Tytler, R., Freeman, B., & Roberts, K. (2013). STEM: Country comparisons: International comparisons of science, technology, engineering and mathematics (STEM) education. Final report. Melbourne, Australia: ACOLA.

Masters, G. N. (2017). How well are we learning from NAPLAN? Teacher, 27 April 2017. https://www.teachermagazine.com.au/columnists/geoff-masters/how-well-are-we-learning-from-naplan

Marsh, J., Kumpulainen, K., Nisha, B., Velicu, A., Blum-Ross, A., Hyatt, D., & Thorsteinsson, G. (2017). Makerspaces in the early years: A literature review. D4 Published development or research report or study.

Marsh, J., Wood, E., Chesworth, L., Nisha, B., Nutbrown, B., & Olney, B. (2019). Makerspaces in early childhood education: Principles of pedagogy and practice. *Mind, Culture, and Activity, 26*(3), 221–233. https://doi.org/10.1080/10749039.2019.1655651

Martin, L. (2015). The promise of the maker movement for education. *Journal of Pre-College Engineering Education Research (J-PEER), 5*(1). https://doi.org/10.7771/2157-9288.1099

Martinez, S. L., & Stager, G. (2019). *Invent to learn: Making, tinkering, and engineering in the classroom*. Constructing Modern Knowledge Press.

Maslow, A. H. (1954). In *Motivation and personality* (3rd ed.) Harper & Row.

Maslow, A. H. (1943). A theory of human motivation. *Psychological Review, 50*(4), 370–396. https://doi.org/10.1037/h0054346

McIlvenny, L. (2019). Transversal competencies in the Australian curriculum. *Access, 33*(2), 6–13.

McIlvenny, L., & Sheffield, R. (2019). Are transversal competencies the new black. *Professional Educator, 20*(1), 42–45.

McIlvenny, L. (2018). Transversal Competencies [website]. https://transversalcompetencies.weebly.com

McLeod, S. (2018). Lev Vygotsky's sociocultural theory. *Simply Psychology*. https://www.simplypsychology.org/vygotsky.html

McKinsey Global Institute. (2017). Skill shift: Automation and the future of the workforce. MGI-Skill-ShiftAutomation-and-future-of-the-workforce-May-2018.ashx

Mckay, C., Banks, T. D., & Wallace, S. (2016). Makerspace classrooms: Where technology intersects with problem, project, and place-based design in classroom curriculum. *International Journal of Designs for Learning, 7*(2). https://doi.org/10.14434/ijdl.v7i2.20267

Meyer, L. (2017). Planning and implementing a makerspace in your school: Our experts offer 14 tips for adding a makerspace to any K-12 school. *T H E Journal (technological Horizons in Education), 44*(3), 26–30.

Ministry of Human Resource Development. (2016). National policy on education. http://www.nuepa.org/New/download/NEP2016/ReportNEP.pdf

Modi, N. (2017). Science, technology innovation keys to India's progress. Business Standard. https://www.business-standard.com/article/news-ani/science-technology-and-innovation-keys-to-india-s-progress-pm-modi-117071900779_1.html

Montessori, M. (1965). In *Dr Montessori's own handbook*. LB Education: Theory and practice of education.

Moore, B. N., & Parker, R. (1994). *Critical thinking. Mountain view*. Mayfield.

Moorefirld-Lang, H. (2015). Change in the making: Makerspaces and the ever-changing landscape of libraries. Association for Educational Communications and Technology. https://www.makersempire.com/wp-content/uploads/2018/02/Change-in-the-Making-Makerspaces-and-the-Ever-Changing-Landscape-of-Libraries-Moorefield-Lang-15.pdf

Meijer, C. (2015). Maakplaatsen. Vestigingsvoorwaarden van fablabs & makerspaces in Nederland. https://docplayer.nl/41135834-Maakplaatsen-vestigingsvoorwaarden-van-fablabsmakerspaces-in-nederland.html

Nagel, D. (2018). Creativity, critical thinking top benefits of makerspaces. *The Journal (Technological Horizons In Education), 45*(2), 34. http://reports.weforum.org/future-of-jobs-2016/chapter-1-the-future-of-jobs-and-skills/

Namka, L. (2014). In *Lesson plans for teaching resilience to children*. Talk, Trust & Feel Therapeutics.

National Council of Educational Research and Training. (2018). Science Exhibition. http://www.ncert.nic.in/programmes/science_exibition/index_science.html

National Education Association. (2019). Preparing 21st century students for a global society: an educator's guide to the "Four Cs". http://www.nea.org/assets/docs/A-Guide-to-Four-Cs.pdf

National Innovation Foundation—India (2018) Inspire Awards—MANAK. http://nif.org.in/inspire-awards

National Research Council. (2012). *A framework for K-12 science education: Practices, crosscutting concepts, and core ideas*. The National Academies Press.

National Science Foundation—STEM Education Data. (2020). US Bureau of Labour Statistics. https://www.nsf.gov/nsb/sei/edTool/data/workforce-03.html

National University of Educational Planning and Administration. (2015). School Education in India. National University of Educational Planning and Administration. http://www.dise.in/Downloads/Publications/Documents/U-DISE-SchoolEducationInIndia-2014-15.pdf

Neve, C. (1986). High lea Ling jumps show potency of brain-based instruction. *Phi Delta Kappan, 68*, 143–148.

Noonoo, S. (2017). Build a makerspace that engages students across every STEAM subject. https://thejournal.com/Articles/2017/02/09/Build-a-Makerspace-ThatEngages-Students-Across-Every-STEAM-Subject

Novak, S. (2019). The effects of a makerspace curriculum on the 4C's in education. Graduate Research Papers. (pp. 942). https://scholarworks.uni.edu/grp/942

Office of the Chief Scientist. (2014). Australia's STEM workforce: A survey of employers. https://www.chiefscientist.gov.au/wp-content/uploads/DAE_OCS-Australias-STEM-Workforce_FINAL-REPORT.pdf

Oliver, K. M. (2016). Professional development considerations for makerspace leaders, part one: Addressing "what?" and "why?" *TechTrends, 60*(2), 160–166. https://ierome.stiohns.edu/login?url=https://search.ebscohost.com/login.aspx7direct=true&db=eric&AN=EJ1094199&login.asp&site=ehost-live
Organisation for Economic Cooperation and Development [OECD]. (2012). What is equity in education? In Education at a Glance Highlights. OECD Publishing. https://doi.org/10.1787/eag_highlights-2012-29-en
Organisation for Economic Co-Operation and Development. (2009). Education at a Glance 2009. http://www.oecd.org/education/skills-beyond-school/43636332.pdf
Organisation for Economic Co-Operation and Development (OECD). (2016). Indonesia. In *OECD Science, technology and innovation outlook 2016*. OECD Publishing. https://doi.org/10.1787/sti_in_outlook-2016-66-en
Organisation for Economic Co-Operation and Development (n.d.) Australia Student Performance (PISA 2015). http://gpseducation.oecd.org/CountryProfile?plotter=h5&primaryCountry=AUS&treshold=10&topic=PI
Organisation for Economic Co-operation and Development (OECD). (2005). The Definition and Selection of Key Competencies Executive Summary. https://www.oecd.org/pisa/35070367.pdf
Organisation for Economic Co-operation and Development (OECD). (2014, November). The OECD skills strategy and the education agenda for development. *International Journal of Educational Development,39*, 40–48. https://doi.org/10.1016/j.ijedudev.2014.08.008
Organisation for Economic Co-operation and Development (OECD). (2012). OECD launches skills strategy to boost jobs and growth. http://www.oecd.org/education/oecdlaunchesskillsstrategytoboostjobsandgrowth.htm
Organisation for Economic Co-operation and Development (OECD). (2012). "PISA 2012 Results: Creative problem solving". In *Students' skills in tackling real-life problems* (pp. 124). http://www.oecd.org/pisa/keyfindings/pisa-2012-results-volume-v.htm
Organisation for Economic Co-operation and Development (OECD). (2014). Skills beyond school: Synthesis report, OECD Reviews of Vocational Education and Training, OECD Publishing.
Organisation for European Economic Cooperation and Development. (2001). What schools for the future, schooling for tomorrow, OECD Publishing.https://doi.org/10.1787/9789264195004-en
Organisation for European Cooperation and Development. (2017). PISA 2015 Assessment and analytical framework science, reading, mathematic, financial literacy and collaborative problem solving. https://www.oecd.org/publications/pisa-2015-assessment-and-analytical-framework-9789264281820-en.htm
Papert, S., & Harel, I. (1991). *Constructionism: Research reports and essays 1985–1990*. Epistemology and Learning Research Group, The Media Lab Massachusetts Institute of Technology.
Pantiwati, Y. (2013). Authentic assessment for improving cognitive skill, critical creative thinking and meta-cognitive awareness. *Journal of Education and Practice 4*(14). https://iiste.org/Journals/index.php/JEP/article/download/6790/6903
Partnership for 21st Century Skills. (2009). P21 Framework Definitions. https://files.eric.ed.gov/fulltext/ED519462.pdf
Partnership for 21st Century Skills. (2011). Creativity and Innovation. https://p21.org/index.php?option=com_content&task=vie w&id=262&Itemid=120
Partnership for 21st Century Learning. (2019). Framework for 21st century learning definitions. Battelle for Kids. https://static.battelleforkids.org/documents/p21/P21_Framework_DefinitionsBFK.pdfhttps://easdinnovativelearning.weebly.com/uploads/7/7/1/6/77162025/p21_framework_definitions_new_logo_2015.pdf
Pascarella, E., & Terenzini, P. (1991). *How college affects students: Findings and insights from twenty years of research*. Jossey Bass.
Paul, R. W. (1995). Critical thinking: How to prepare students for a rapidly changing world. Foundation for Critical Thinking.
Paul, R., & Elder, L. (2006). Critical thinking: The nature of critical and creative thought. *Journal of Developmental Education., 30*(2), 34–35.

Paul, R., & Elder, L. (2012). In *Critical and creative thinking*. Rowman & Littlefield Publishing Group Inc.

Pepper, D. (2011). Assessing key competences across the curriculum—and Europe. *European Journal of Education Special Issue: Key Competences in Europe, 46*(3), 335–353.

Peppler, K. A., & Bender, S. (2013). Maker movement spreads innovation one project at a time. *Phi Delta Kappan, 95*(3), 22–27. https://doi.org/10.1177/003172171309500306

Permatasari, C. P., Yerizon, Y., Arnawa, I., & Musdi, E. (2020). Improving students' problem-solving ability through learning tools based on problem based learning to conference on mathematics and mathematics. *Education Journal of Physics: Conference Series 1554*. https://doi.org/10.1088/1742-6596/1554/1/012017 https://www.researchgate.net/publication/341759089_Improving_Students'_Problem-Solving_Ability_through_Learning_Tools_Based_on_Problem_Based_Learning

Piaget, J. (1973). To understand is to invent: The future of education. Grossman Publishers New York.

Pianta, R. C., & Walsh, D. J. (1998). Applying the construct of resilience in schools: Cautions from a developmental systems perspective. *School Psychology Review, 27*(3), 407.

Plomp, T. (2013). Preparing education for the information society: The need for new knowledge and skills. *International Journal of Social Media and Interactive Learning, 1*(1), 3–18. https://doi.org/10.01504/IJSMILE.2013.05161

Poysa-Tarhonen, J., Care, E., Awwal, N., & Hakkinen, P. (2018). Pair interactions in online assessments of collaborative problem solving: case-based portraits. *Research and Practice in Technology Enhanced Learning* (Vol. 1). Springer. https://doi.org/10.1186/s41039-018-0079-7

Press Information Bureau (2014, November 27) GDP in science research. Retrieved from http://pib.nic.in/newsite/PrintRelease.aspx?relid=112011

Price-Mitchell, P. M. (2015). 8 pathways to every student's success. Edutopia. Social and Emotional Learning. Retrieved from https://www.edutopia.org/blog/8-pathways-every-students-success-marilyn-price-mitchell.

Price-Mitchell, M. (2015). Does your classroom cultivate student resilience? Retrieved from https://www.edutopia.org/blog/8-pathways-cultivate-student-resilience-marilyn-price-mitchell.

Quinlan, O. (2015). Young digital makers—Surveying attitudes and opportunities for digital creativity across the UK. https://media.nesta.org.uk/documents/youngdigmakers.pdf

Ravipati, S. (2017). Maryland public libraries launches virtual makerspace. *The Journal*. https://thejournal.com/articles/2017/01/09/maryland-public-libraries-launch-virtual-makerspace.aspx

Rendina, D. (2018). *5 Reasons Makerspaces Belong in School Libraries*. https://ideas.demco.com/blog/5-reasons-makerspaces-belong-in-school-libraries/

Reyes, J. A., Elias, M. J., Parker, S. J., & Rosenblatt, J. L. (2013). Promoting educational equity in disadvantaged youth: The role of resilience and social-emotional learning. In *Handbook of resilience in children* (pp. 349–370). Springer.

Robinson, K. (2011). Ken Robinson says schools kill creativity. speech. TED Talks. TED Talks. Web. 16 May 2011. http://www.ted.com/talks/ken_robinson_says_schools_kill_creativity.html

Rodgers, C. (2002). Seeing student learning: Teacher change and the role of reflection. *Harvard Educational Review., 72*(2), 846.

Rosa, P., Ferretti, F., Guimarães Pereira, A., Panella, F., & Wanner, M. (2017). Overview of the maker movement in the European Union. JRC Technical Report. Publications Office of the European Union. https://ec.europa.eu/jrc/en/publication/overview-maker-movement-european-union

Rosa, P., Guimarães Pereira, A., & Ferretti, F. (2018). Futures of work: Perspectives from the maker movement (Text No. JRC Technical Report). Publications Office of the European.

Rosin, H. (2014). The overprotected kid. https://www.theatlantic.com/magazine/archive/2014/04/hey-parents-leave-those-kids-alone/358631/

Rand Corporation. (2012). Teaching and learning 21st century skills. http://asiasociety.org/files/ramd-1012repot.pdf.

Robinson K (2010) Changing education paradigms, TED talk. Retrieved from https://www.ted.com/talks/sir_ken_robinson_changing_education_paradigms.

Rudd, R. D., Baker, M. T., Hoover, T. S., & Gregg, A. (1999). Learning styles and critical thinking abilities of college of agriculture students at the University of Florida. In *Proceedings of the 48th Annual Southern Region Agricultural Education Research Meeting* (pp. 123–134).

Rudd, R., & Pomeroy, C. (2001). Relationships between critical and creative thinking. *Journal of Southern Agricultural Education Research 51*(1). https://www.researchgate.net/profile/Matt_Baker4/publication/251671840_Relationships_between_Critical_and_Creative_Thinking/links/53f354370cf2da8797446876/Relationships-between-Critical-and-Creative-Thinking.pdf

Rutter, M., Maughan, B., Mortimore, P., Ouston, J., & Smith, A. (1979). *15,000 hours: Secondary schools and their effects on children*. Open Books.

Ruutmann, T. (2019). Development of critical thinking and reflection. https://doi.org/10.1007/978-3-030-11935-5_85

Ryan, T. (2014). Thinkers keys. HeadFirst Publication. https://books.google.com.au/books/about/Thinkers_Keys.html?id=Yz1PBAAAQBAJ&printsec=frontcover&source=kp_read_button&redir_esc=y#v=onepage&q&f=false

Sahlberg, P. (2010). The secret to Finland's success: Educating teachers. Research brief stanford center for opportunity policy in education research brief centre for international mobility and cooperation, September 2010. https://eric.ed.gov/?id=ED533000 https://www.researchgate.net/publication/326713638_RETHINKING_OUR_EDUCATION_TO_FACE_THE_NEW_INDUSTRY_ERA [accessed Feb 12 2021]

Sarkadi, A. R. C., & Rahmawati, Y. (2020). Imrpoved learning design for pre-service teacher in a character education course. *Universal Journal of Education Research, 8*(1), 212–224. https://doi.org/10.13189/ujer.2020.080126

Salmon, G. (2019). May the fourth be with you: Creating education 4.0. *Journal of Learning for Development, 6*(1), 95–115. https://files.eric.ed.gov/fulltext/EJ1222907.pdf

Sampurno, P. J., Sari, Y. A., & Wijaya, A. D. (2015). Integrating STEM (Science, Technology, Engineering, Mathematics) and disaster (STEM-D) education for building students' disaster literacy. *International Journal of Learning and Teaching, 1*(1), 73–76. https://doi.org/10.18178/ijlt.1.1.73-76

Sarangapani, P. M. (2017). Three challenges facing Indian school science education. http://www.esocialsciences.org/eSS_essay/Science_education/Three%20Challenges%20Facing%20India_Sarangapani.pdf

Santos, E., & Benneworth, P. (2019). Makerspace for skills development in the industry 4.0 era. *Brazilian Journal of Operations and Production Management, 16*(2), 303–315. https://doi.org/10.14488/BJOPM.2019.v16.n2.a11

Scriven, M. (1976). *Reasoning*. McGraw-Hill.

Sejnost, R. (2009). Tools for teaching in the block. In *Tools for teaching in the block* (pp. 23–66). Corwin Press. https://doi.org/10.4135/9781452219097.n2

Seligman, M. E., & Csikszentmihalyi, M. (2000). Positive psychology: An introduction. *55*(1), 5. American Psychological Association.

Septiani, A., & Rustaman, N. Y. (2017). Implementation of performance assessment in STEM (Science, Technology, Engineering, Mathematics) education to detect science process skill. *Journal of Physics: Conference Series, 9*, 3–10. https://doi.org/10.1088/1742-6596/812/1/012052/pdf

Sheffield, R., & Blackley, S. (2015). What has STEM ever done for us? In *Science teachers association of Western Australia* (Ed), Future Science Conference held in Curtin University, Perth, WA, 4th December 2015.

Sheffield, R., & Blackley, S. (2015) E for enquiry in STEM education. In *Eight International Conference on Science, Mathematics and Technology Education* (8th SMTE). Jakarta, Indonesia 21–24 November 2015.

Sheridan, K., Halverson, E. R., Litts, B., Brahms, L., Jacobs-Priebe, L., & Owens, T. (2014) Learning in the making: A comparative case study of three makerspaces. *Harvard Educational Review, 84*(4), 505−531.https://doi.org/10.17763/haer.84.4.brr34733723j648u

Smay, D., & Walker, C. (2015). Makerspaces: A creative approach to education. *Teacher Librarian, 42*(4), 39–43.
Smith, A., Hielscher, S., Dickel, S., Soderberg, J., & Oost, E. V. (2013). Grassroots digital fabrication and makerspaces: Rreconfiguring, relocating and recalibrating innovation?.
Smith, B. W., Dalen, J., Wiggins, K., Tooley, E., Christopher, P., & Bernard, J. (2008). The brief resilience scale: Assessing the ability to bounce back. *International Journal of Behavioral Medicine., 15*, 194–200. https://doi.org/10.1080/10705500802222972
Speak Up Research Project for Digital Learning. (n.d.). https://tomorrow.org/speakup/
Stanford Research Institute [SRI] International (n.d.) 21st Century Learning Design (21CLD). https://www.sri.com/work/projects/21st-century-learning-design-21cld
Stall, N. N., & Stahl, R. J. (1991). We can agree after all: Achieving a consensus for a critical thinking component of a gifted program using the Delphi technique. *Roeper Review, 14*(2), 79–88.
Sternberg, R. J. (2005). Successful intelligence in the classroom. http://arowepbworks.com/f/Sternberg_on_intelligences.pdf
Sucha, V., Mair, D., & Sienkiewitz, M. (Eds.) (forthcoming) In *Science and evidence in the policy ecosystem handbook*. Elsevier.
Surowiecki, J. (2005). *The wisdom of crowds*. Anchor, Print.
Surviving and thriving in the Fourth Industrial Revolution: Digital skills for education and society. https://www.researchgate.net/publication/332039429_Surviving_and_thriving_in_the_Fourth_Industrial_Revolution_Digital_skills_for_education_and_society [Accessed Dec 31 2020].
Tan, S., Al-Jumeily, D., Mustafina, J., Hussain, A., Broderick, A., & Forsyth, H. (2018). Rethinking our education to face the new industry era. In *10th International Conference on Education and New Learning Technologies*.
Taylor, B. (2016). Evaluating the benefit of the maker movement in K-12 STEM education. *Electronic International Journal of Education, Arts, and Science (EIJEAS), 2.* http://www.eijeas.com/index.php/EIJEAS/article/view/7243UNESCO. (2017)
Taylor, N., Hurley, U., & Connolly, P. (2016). Making community: the wider role of makerspaces in public life. Conference: CHI 2016. https://doi.org/10.1145/2858036.2858073
The Fab Charter (October 2012). http://fab.cba.mit.edu/about/charter/
The Partnership for 21st Skills. (2020). P21 framework definitions. Retrieved from https://files.eric.ed.gov/fulltext/ED519462.pdf
Thomas, A., & Thorne, G. (2009). How to increase higher order thinking. Center for Development and Learning. Retrieved Dec. 7, 2009, from http://www.cdl.org/resource-library/articles/HOT.php?type=subject&id=18
Thomson, S., Wernert, N., O'Grady, E. J., & Rodrigues, S. (2017). TIMSS 2015: Reporting Australia's results. https://research.acer.edu.au/cgi/viewcontent.cgi?article=1002&context=timss_2015
The Book of Life. (2018). Kintsugi. https://www.theschooloflife.com/thebookoflife/kintsugi-2/
Timms, M., Moyle, K., Weldon, P., & Mitchell, P. (2018). Challenges in STEM learning in Australian schools. Policy Insights (Vol. 7). ACER.
Tinzmann, M. B., Jones, B. F., Fennimore, T. F., Bakker, J., Fine, C., & Pierce, J. (1990). What is the collaborative classroom?: New learning and thinking curricula require collaboration NCREL. http://methodenpool.uni-koeln.de/koopunterricht/The%20Collaborative%20Classroom.htm
Tishman, S., Perkins, D. N., & Jay, E. (1995). The thinking classroom: Learning and teaching in a culture of thinking (Needham Heights).
Tocino-Smith, J. (2019). Teaching resilience in schools and fostering resilient learners. https://positivepsychology.com/teaching-resilience/
Torrance, E. P. (1966). Rationale of the Torrance tests of creative thinking ability. In E. P. Torrance & W. F. White (Eds.), 1969 Issues and advances in education psychology. F. E. Peacock.
Torrance, C. A., & Nagel, D. (2018). Creativity, critical thinking top benefits of makerspaces. *The Journal (technological Horizons in Education), 45*(2), 34.

Tsekeris, C. (2019). Surviving and thriving in the fourth industrial revolution: Digital skills for education and society. *EKT, Homo Virtualis, 2*(1), 34–42. https://doi.org/10.12681/homvir.20192
Turk (n.d.) Setting up a collaborative makerspace. Zspace. https://zspace.com/blog/setting-up-a-makerspace
UNICEF. (2018). Catalysing transformational change in school education. Child Rights. Published December 2018, (pp. 5–7).
Union, Luxembourg (n.d) ISBN 978–92-79–89734–4, https://doi.org/10.2760/96812. https://ec.europa.eu/jrc/en/publication/overviewmaker-movement-european-union
United Nations. (2020). United nations sustainable development goals. Goal 4: Quality Education. Retrieved from https://www.un.org/sustainabledevelopment/education/
United Nations Educational Scientific and Cultural Organisation (UNESCO). (2015). Asia-Pacific Education Research Institutes Network (ERI-Net) regional study on transversal competencies in education policy and practice (Phase I): Regional Synthesis Report. https://unesdoc.unesco.org/ark:/48223/ pf0000231907
United Nations Educational Scientific and Cultural Organisation. (2017) Cracking the code: Girls' and women's education in science, technology, engineering and mathematics (STEM), UNESCO.
United Nations Educational, Scientific and Cultural Organisation [UNESCO]. (2017). Digital skills for life and work. Drafted by the Broadband Commission for Sustainable Development's Working Group on Education. http://www.skillsforemployment.org/KSP/en/Details/?dn=WCMSTEST4_193296
University of Southern California (n.d.) Makerspaces: Why libraries are the perfect fit [Blog]. https://librarysciencedegree.usc.edu/blog/makerspaces-in-libraries/
Urban Libraries Council (n.d.) Libraries are an ideal setting for Makerspaces. https://www.urbanlibraries.org/member-resources/makerspaces-in-libraries
USC Rossier (n.d) How maker education is impacting student cognition. USC Rossier School of Education https://rossieronline.usc.edu/maker-education/student-learning/.
Velasquez, J. (2011). MakerSpaces: New tradition in context. International Federation of Library Associations. https://www.ifla.org/node/9625
Venville, G., Adey, P., Larkin, S., Robertson, A., & Hammersmith, F. (2003). Fostering thinking through science in the early years of schooling. *International Journal of Science Education, 25*(11), 1313–1331. https://doi.org/10.1080/0950069032000052090
Verma, G. (2019). Jugaad thinking: Contextualised innovative thinking in India through science, technology, engineering and mathematics (STEM) Education?. In R. Koul, G. Verma, & V. Nargund-Joshi (Eds.), *Science education in India*. Springer. https://doi.org/10.1007/978-981-13-9593-2_11
Voigt, C., Suero Montero, C., & Menichinelli, M. (2016). An empirically informed taxonomy for the maker movement. In F. Bagnoli et al. (Eds.), *Internet science*. INSCI 2016. Lecture Notes in Computer Science (Vol. 9934). Springer.
Vongkulluksn, V. W., Matewos, A. M., Sinatra, G. M., & Marsh, J. A. (2018). Motivational factors in makerspaces: A mixed methods study of elementary school students' situational interest, self-efficacy, and achievement emotions. *International Journal of STEM Education, 5*(1), 1–19. https://doi.org/10.1186/s40594-018-0129-0
Vongkulluksn et al (2018); Nagel (2018); Clapp et al (2017); Sheridan et al, (2014) *An Educator's Guide to the "Four Cs.* (n.d.). http://www.nea.org/tools/52217.htm
Voogt, J., & Roblin, N. P. (2012). A comparative analysis of international frameworks for 21st century competences: Implications for national curriculum policies. *Journal of Curriculum Studies, 44*(3), 299–321.
Vuorikari, R. (2018). Innovating professional development in compulsory education—Examples and cases of emerging practices for teacher PD. In JRC Technical Report. Publications Office of the European Union. https://doi.org/10.2760/734136.
Vuorikari, R., Punie, Y., Carreton, S., & Van den Brande, L. (2016). DigComp 2.0: The digital competence framework for citizens. https://publications.jrc.ec.europa.eu/repository/bitstream/

JRC101254/jrc101254_digcomp/202.0/20the/20digital/20competence/20framework/20for/20citizens./20update/20phase/201.pdf
Vuorikari, R., Ferrari, A., & Punie, Y. (2019). Makerspaces for education and training—Exploring future implications for Europe. Publications Office of the European Union. ISBN 978-92-76-09032-8, JRC117481 https://doi.org/10.2760/946996.
Vygotsky, L. (1997). In *Interaction between learning and development.* W.H. Freeman and Company.
Wang, F., Wang, W., Wilson, S., & Ahmed, N. (2016). The state of library makerspaces. *International Journal of Librarianship, 1*(1), 2–16.
We Are Water Foundation India. (2021). India's challenges. Retrieved from https://www.wearewater.org/en-IN/india-s-challenges_285111.
Wechsler, S. M. (2003). Assessing Brazilian creativity with Torrance Tests. In M. S. Stein (Ed.), (org) *Creativity global correspondents.* Winslow Press.
White House. (2014). Presidential proclamation—National day of making. https://obamawhitehouse.archives.gov/the-press-office/2014/06/17/presidential-proclamation-national-day-making-2014
Wiederhold, C., & Kagan, S. (1991). The question matrix: Cooperative learning and critical thinking: Resources for teachers.
Willingham, T. (2017). *Library makerspaces: The complete guide.* Rowman & Littlefield.
World Economic Forum (2016) The Global Human Capital Report 2016. https://www.weforum.org/reports/thehuman-capital-report-2016
World Economic Forum. (2017). The Global Human Capital Report 2017. https://weforum.ent.box.com/s/dari4dktg4jt2g9xo2o5pksjpatvawdb
World Economic Forum (2021) Education 4.0.
https://www.weforum.org/projects/learning-4-0?fbclid=IwAR2My1baMbT5nw8-kpzBTUoSju_aTmFFP2SIETXR2OHU2CBCL8X791PVVa4
Worsley, M., & Blikstein, P. (2018). A multimodal analysis of making. *International Journal of Artificial Intelligence in Education, 28*(3), 385–419. https://doi.org/10.1007/s40593-017-0160-1
Youngmoo, E. K., Edouard, K. (2018). National study of makerspaces. Slides. https://drexel.edu/~/media/Files/excite/Making%20Culture%20%20SXSW%20EDU%202019.ashx?la=en
Youngmoo, E. K., Edouard, K., Alderfer, K., & Smith, B. (2018). Making culture: A national study of Ed makerspaces. ExCITe Center. https://drexel.edu/~/media/Files/excite/making-culture-full-report.ashx
Zervas, P., & Sampson, D. (2016). Supporting reflective lesson planning based on inquiry analytics for facilitating students' problem-solving competence development: The inspiring science education tools. In *Lecture Notes in Educational Technology*, (pp. 91–114).

The manufacturer's authorised representative in the EU is Springer Nature Customer Service Centre GmbH, Europaplatz 3, 69115 Heidelberg, Germany. If you have any concerns regarding our products, please contact ProductSafety@springernature.com

Printed and bound by CPI Group (UK) Ltd, Croydon, CR0 4YY
30/09/2026
02225317-0003